PRINCE

of

ALES

The History of

Brewing in Wales

SOLID BUSINESS: one of the earlier Swansea breweries, Andrew and Crowhurst's Orange Street Brewery, pictured in the 1860s (Swansea Archive)

PRINCE

of

ALES

The History of
Brewing in Wales

BRIAN GLOVER

ALAN SUTTON

First published in the United Kingdom in 1993 by
Alan Sutton Publishing Ltd · Phoenix Mill · Far Thrupp · Stroud · Gloucestershire

First published in the United States of America in 1993 by
Alan Sutton Publishing Inc. · 83 Washington Street · Dover · NH 03820

British Library Cataloguing in Publication Data

Glover, Brian
  Prince of Ales: History of Brewing in Wales
  I. Title
  663.309429

ISBN 0–7509–0331–7

Library of Congress Cataloging in Publication Data applied for

Welsh Ales poster on the front cover courtesy of Clwyd Archives.

Typeset in 11/13 Perpetua.
Typesetting and origination by
Alan Sutton Publishing Limited.
Printed in Great Britain by
The Bath Press, Bath, Avon.

CONTENTS

PREFACE

This book sprang out of the realization that there was virtually nothing written on the history of brewing in Wales. And what was left of that heady heritage was fast disappearing. It has taken four years to piece together this broken barrel.

It would be nice to think that most of the research had been conducted in pubs, but in reality it has meant long, dry hours in libraries and archives. I would like to thank their staff for putting up with my many strange requests. I am also grateful for the co-operation of the few surviving breweries, and in particular to Welsh Brewers for sponsoring this book. I would especially like to thank Vernon Howells of Cardiff for his early enthusiasm and help, and Nick Redman, Whitbread's company archivist. His book *Whitbread in South Wales* proved invaluable. The Brewery History Society's book *Where Have All the Breweries Gone?* also proved useful for checking the directory of Welsh breweries at the back of this book.

I am conscious that there are still far too many gaps in my knowledge, and so I would welcome any additional information. I am also interested in discovering further photographs and publicity material from Welsh breweries. If you can help, please write to me at 26 Windsor Avenue, Radyr, Cardiff CF4 8BY.

BRIAN GLOVER

FIRST GLASS: 'The taste of the beer, its fine white lather, its brassed-bright depths, the sudden world through the wet-brown walls of glass, the tilted rush to the lips and the slow swallowing down to the lapping belly, the salt on the tongue, the foam at the corners.'

DYLAN THOMAS

WELSH ALE

Long before the English could boast about their Burton beer and London porter, Welsh ale was famous – and much valued by men of wealth from noblemen and bishops to kings. When King Ine of Wessex drew up a set of laws between AD 690 and 693, one of them referred to payments in kind in return for land. The law ruled that for every ten hides, the food rent should be:

> Ten vats of honey, 300 loaves, 12 ambers of Welsh ale, 30 ambers of clear ale, two full grown cows or ten wethers, ten geese, 20 hens, ten cheeses, a full amber of butter, five salmon, 20 pounds of fodder and 100 eels.

Land did not come cheap. There may have been scope to haggle over the length of eels, but there was no chance of wriggling out of a clear commitment to hand over 'twelve ambers of Welsh ale'.

Almost 500 years later Welsh ale was still much sought after, though the eels seem to have been forgotten. A twelfth-century agreement between Ceolred, Abbot of Medeshamstede (Peterborough) and his tenant Wulfred, granted Wulfred the church-owned estate at Sempringham in return every year for a horse, 30 shillings and one day's food rent including 15 mittan of clear ale, five mittan of Welsh ale and 15 sesters of mild ale.

Welsh ale did not need to come from Wales. It was a well recognized recipe, a brewing style out on its own. The name implies that Welsh ale was a pre-Saxon drink, driven west by the invaders and then welcomed back into England as its heady flavour was appreciated by the new masters.

When King Offa granted the lands at Westbury and Stanbury to the church of Worcester, he accepted in return, among other payments, 'two tunne full of clear ale' and 'a cumbe (16 quarts) of Welsh ale'.

Records of this distant period regularly divide ale into three types – clear, mild and Welsh. The latter was evidently neither clear nor mild, and much more valuable than the other two. It was a strong, heavy brew laced with expensive spices like cinnamon, cloves and ginger.

Denewulf, Bishop of Winchester, when leasing an estate from King Edward in the year 901, had to pay 12 sesters of 'sweet Welsh ale' as part of his annual rent. This was more likely flavoured with honey. Some land agreements even specified that of the casks of Welsh ale 'quorum unum fit melle dulcoratum' (i.e. one should be sweetened with honey).

Mead – fermented honey – was the most highly-prized Celtic drink. According to one Welsh law the dimensions of a cask of mead must be nine palms in height and wide enough 'to serve the king and one of his counsellors for a bathing tub'. The bees would need to be very busy, a hive of activity, to fill it.

Another Welsh law said there were two types of Welsh ale – common ale and spiced ale – and fixed their value in relation to mead. 'If a farmer hath no mead [to pay part of his rent] he shall pay two casks of spiced ale or four casks of common ale for one cask of mead.'

John Bickerdyke in his nineteenth-century book, *The Curiosities of Ale and Beer,* said:

Welsh ales were in Saxon times well known and highly esteemed. In the laws of Hywel Dda (Howell the Good) two kinds of ale are mentioned – Bragawd or Bragot, which was paid as tribute to the king by a free township, and Cwrwf, which was more common and was paid by the servile township in cases where the former kind ran short.

Cwrwf, or cwrw as it later became, was the standard ale of the Welsh for everyday drinking. The Welsh ale mentioned in England would be the spiced ale. As Bickerdyke concludes: 'It may be hence gathered that in early times the highly-flavoured Bragawd was held in greater estimation than the Cwrwf (common ale).'

A medieval recipe for bragawd involves mixing 'fyne wort' (the liquid extracted from mashing malted barley or other cereals in hot water) with honey, cinnamon, cloves, ginger, pepper and galingale (a type of rush).

The thought of sampling such a strange brew might make seasoned beer-drinkers squirm, but remember that when the hop was first introduced into Britain in the fourteenth century many objected to this 'pernicious weed' which imparted an odd bitter flavour to their beer. Henry VIII even banned the hop for a time.

In fact the authorities' insistence on ale being brewed only from malt, yeast and water – because too many unscrupulous brewers were adulterating their brews with unsavoury ingredients – led to the demise of spiced Welsh ale.

Wales's early claim to brewing fame became just a distant hangover.

HOME-BREWED BEER

Early land agreements in Britain show one thing clearly – all households were expected to brew beer. It was as basic a part of everyday life as cooking, and as a domestic chore was at first regarded as women's work.

Ale wives were a familiar part of early Britain. There is a description in the Chester Miracle Plays of a fraudulent ale wife or brewster (female brewer) 'who kept no true measure, deceiving many'. This serious crime ensured her passage to hell.

The long fourteenth-century allegorical poem, *The Vision of Piers Plowman*, contains many references to ale wives:

> My wife was a webbe [weaver] and woollen cloth made
> I bought her barley malt, she brewed it to sell;
> Penny ale and pudding ale she poured together
> For laborers and for low folk.
> The best ale lay in my bower or in my bed chamber,
> And whoso tasted thereof, bought it thereafter,
> A gallon for a groat – God wot, no less.

Penny ale (penny a gallon not a pint!) was the weaker common ale, sometimes known as table or small beer. Pudding ale was the stronger brew, often with added herbs though these were less exotic than in Welsh ale. These could be specifically added at the request of the customer. When Plowman asked one ale wife what she had in store she replied, 'I have pepper and peonyseeds and a pound of garlic, and a farthingsworth of fennelseed for fasting days.' Rosemary was another favourite. These flavourings may have been used to mask the taste of sour beer.

The exception to this domestic business was the monasteries where monks brewed on a surprisingly large scale. The Friary at Carmarthen, for instance, had a brewhouse between the kitchen and the buttery. The monks needed to brew larger quantities not only for their own communities but also for the many travellers who stayed there.

They are credited with introducing the system of crosses on barrels to denote the strength of different brews. Some must have spent too long over their casks since laws were passed that any brother who slurred his singing during the psalms was to lose his supper.

Henry VIII's large figure brought this early pattern of brewing to a sudden end. In 1536 he dissolved the monasteries and four years later ruled that it was unseemly for women under forty to sell ale or keep alehouses.

But ale was not going to go away. At a time when the purity of water was uncertain and tea and coffee were unknown – and when they did finally arrive were far too expensive for the average family – beer provided a safe daily drink. Though some of the vessels used could be a little worrying.

Thomas Pennant, writing about his *Tours in Wales* of 1778 and 1781, recalled:

> He welcomed us with ale and potent beer to wash down the Coch yr Wden, or hung goat, and the cheese compounded of the milk of cow and sheep. He likewise showed us the ancient family cup made of a bull's scrotum, in which large libations had been made in days of yore.

In what was a largely rural country, home-brewed beer was provided by Welsh farmers for their labourers with virtually every meal. At harvest time liberal amounts would be available for the extra hands to wash away the dry dust of the fields; it was an unwritten part of their contract.

Beer lubricated all the other main social events of the agricultural calendar from threshing and sheep-shearing to milling and barn building. Ploughing matches were notorious for their heavy drinking. Pig-killing in the cold month of January followed a special routine of warmed beer or ale fortified with spirits.

Stronger brews were used at auctions of farms, animals or machinery to ensure lively bidding; the auctioneer often postponing the start of the sale until the afternoon, when

HEADY HARVEST: some Welsh farmers have continued to the present brewing their own beer. Alan Jones of Aberlleine Farm, near Llandysul, holds up a pint of his own brew in 1986, based on a family recipe handed down the generations (Western Mail)

everyone was merry, in order to push the prices up. In 1906 temperance societies moved a resolution to ban beer at farm auctions.

Special ales also played a central role in all the annual celebrations from harvest suppers to Christmas and the New Year. Bidding ales would be produced for weddings. Brian Waters in his book *The Bristol Channel* (1955) recalled that this ceremony persisted for many years on Gower, when sides of beef would be set up in a barn with plenty of space for dancing country reels.

During these jollifications home-brewed beer was sold to the guests (who had already paid a half-crown entrance fee) as in a public house, and on the proceeds of half-crowns, heaves (extra half-crowns from a plate passed round) and pints of home-brew, the young couple were set up for the first years of their married life.

Care had to be taken that the authorities were looking the other way. A farmer in West Wales was fined £12 in 1878 'for having been so regardless or oblivious of the licensing laws as to brew a little beer for the bidding connected with the approaching marriage of his daughter.' Some got round this problem by charging only for the food.

A similar gathering was the cwrw bach feast among poorer families. When a member of the community suffered from illness, accident or other misfortune, folk would gather round and hold a meal with home-brewed beer. The money raised would be given to the affected family.

The process of domestic brewing was quite simple and did not involve large vessels, though it was fairly time consuming. The initial process would take the best part of a day. Cobbett in his work of 1822, *Cottage Economy*, devoted considerable space to brewing.

First a few gallons of water would be heated in a brass pan over a fire and then poured into an open-top wooden barrel or similar vessel. Barley malt, which had been mixed with water into a porridge, was then stirred in and the whole left to cool under a blanket. More water, as required, was heated and poured in. The strength of the beer depends on the relative amounts of water and malt used. The colour of the malt determined the colour of the beer.

After a few hours of cooling and settling, the liquid was carefully drawn from the barrel through a tap

SEALING THE MASH: a farmer and his assistant put the lid on their barrel-shaped mash tun known as a cerwyn in West Wales (Welsh Folk Museum)

at the bottom. In West Wales a spiny branch of furze or gorse was placed in the bottom of the vat by the tap hole to prevent the sediment running out as well. This process then might be repeated with another batch of water but the same grain to produce a weaker brew, the second runnings. Third runnings were not unknown.

This sweet liquid (wort) was then boiled up with hops (and in later years sugar to further increase the strength), strained to remove the hops, and then poured back into the cleaned out barrel. The spent grains were fed to the pigs. After cooling, yeast was added and the beer left to ferment for over a week, excess yeast being skimmed off the top daily. Then the brew was poured into earthenware jars, and a week to ten days later was judged fit for drinking.

Elfyn Scourfield of the Welsh Folk Museum, who gives a more detailed account of this operation in his booklet *Farmhouse Brewing*, says that 'as an alternative other additives, such as tobacco or whisky, may be included in the final process'.

There were cheaper and simpler methods. Some did not bother to buy in hops or use malt. According to Brian Waters:

An old woman who kept a little inn on Gower gathered her hops wild from the hedges, drying them in bags from the kitchen ceiling, and by this economy was able to brew an 18-gallon cask of beer for one shilling and sixpence.

A shilling was paid for 14 lbs of barley, which was sprouted for five days before being boiled and steeped, four pence was spent on balm (yeast) and additionally, and somewhat mysteriously, half-an-ounce of Bristol twist tobacco was added, for colour and flavour, for two pence. A man who had drunk this brew in his youth told me: 'When you'd had four pints of it you were drunk'.

Some farm beer was put to other uses than drinking. Thomas Jenkins of Aber Drychwallt in the Vale of Neath was a nineteenth-century farmer and veterinary surgeon noted for his 'Beer for Calves'. A letter of 1855 urgently demanded: 'I want 15 quarts of the beer against Wednesday the 17 instant. I will send the jugs that day and also 14 quarts for Mr Thomas of Waterton Court. I will pay for the whole together', wrote Leyson Morgan, adding, 'I have given this incomparable beer for 37 years and not lost one calf.'

The other main drink in rural Wales was buttermilk. Cider was also popular in south-east Wales close to the Herefordshire border. Brecknockshire was once damned by a temperance preacher as 'the cider-besotten county'.

Following the demise of the monasteries more taverns had grown up to meet the needs of travellers, but already every market town and village of size had long developed their own inns, where people could meet and socialize over a drop of beer brewed on the premises.

These public houses were often regarded with suspicion by those in authority as possible centres of sedition or at least places of rowdy behaviour. Rules were regularly passed about their conduct. At the same time local lords were reluctant to close them as they could be useful sources of income.

In the Middle Ages those brewing for sale were supposed to pay a sum in kind or money – the 'prise of ale' or prisage. These could be sizable amounts. On the death of Roger Bigod, Lord of Chepstow, it was stated that 'the prisage was £20, because the lord ought to receive from each tavern, at each brewing, 32 bottles of the best beer'.

In 1622 an order in council was sent out to all Justices of the Peace in the English and Welsh counties stating that in order to relieve the high price of corn through the use of barley, brewing had to be restricted. The JPs were ordered 'to suppress all such ale-houses as were not needful for convenience of the people, and to take strict care that in those still permitted the beer and ale brewed were only of moderate strength'.

Common or larger-scale commercial brewers were of little importance at this time, apart from those in London and a handful of other English cities. Certainly there were none in Wales until the late eighteenth century. But there was one industry connected with brewing which was already thriving within the Welsh borders – malting.

The turning of locally-grown barley into malt, the basic ingredient of beer, required a specialized building with steeping tanks, open floors for germinating the grains and a kiln to dry them. Unlike brewing, it was not something you could easily do in your kitchen or in an outbuilding at the back of a pub.

When the first trade directories began to appear in Wales in the early part of the nineteenth century, they reveal that malting was already a well-established business in many rural areas. Carmarthen in 1835 had fourteen maltsters but only one commercial brewer. 'The trade in malt is respectable', commented Slater's directory of 1844.

The story was repeated to a lesser extent in country towns across South Wales. Llandovery had seven maltsters but no common brewers in 1835. There were six maltsters

EARLY INDUSTRY: because of its specialist nature, malting became a commercial enterprise earlier than brewing in Wales. This sketch from Barnard's Noted Breweries of Great Britain and Ireland shows one of the large malting floors of the Cardiff Malting Company at East Moors, Cardiff, around 1890

in Bridgend in the same year. Cowbridge counted eight in 1840, but its first significant brewer did not appear until the late 1850s. Brecon boasted eleven maltsters and two brewers in 1844. Abergavenny listed ten in 1822, but its first brewers did not surface until twenty years later.

The picture was similar in North Wales. Caernarfon supported ten maltsters in 1835, but never developed a significant brewery. Pwllheli had the same number in 1828. Mold listed nine in 1822. Malting there was described as an 'important business'.

In parts of central Wales the contrast was even greater. Welshpool had eighteen maltsters in 1828; Newtown sixteen in 1835. Neither had any commercial brewers at the time. Charles Humphreys, writing about the small town of Llanfair Caereinion, eight miles from Welshpool, in the mid-nineteenth century, commented:

> The common drink in those days was either buttermilk or beer. Tea or coffee was only drunk on special occasions. Not a pint of the beer drunk was imported from outside. Every publican brewed his beer in his own back-kitchen.
>
> Now to supply these public houses, and there were about a dozen of them, and also practically every farm house in the district brewed their own beer, there had to be a big malting business. There was a malt house in Wesley Street, another on the Mount Road, another near the Poplars, and another large one at the Upper Tanhouse as well as others in the parish. There was always stationed at Llanfair a government excise officer who was fully occupied in testing and supervising the malting and brewing.
>
> Large loads of barley were brought into the town to be converted into malt. As good malting barley could not be grown in the uplands of Montgomeryshire and Merionethshire, the maltsters of Llanfair also supplied the requirements of the thirsty people of the Machynlleth and Dolgelley districts. Llanfair in those days was a miniature Burton-on-Trent.

This area was the only part of Wales to develop hop growing on any scale. According to a parliamentary return of 1833, 130 acres of land in the Principality were devoted to hop fields. The overwhelming bulk of this cultivation – 129 acres – was in mid Wales.

These country maltings would not have been huge, like the vast buildings erected around the main malting towns in Britain, such as Newark in Nottinghamshire and Ware in Hertfordshire. The amount of barley grown in Wales could not compare with the fertile plains of eastern England, but it was still significant. In 1871 barley was grown on 182,000 acres. Only with the growth of the common brewers, who bought in their larger amounts from further afield, did the crop in Wales dwindle. It was halved to 92,000 acres by 1913.

As a local industry in a largely rural economy, malting was important in cereal-growing areas. It was certainly a much bigger business than commercial brewing in the early to mid-1800s. Some of these maltsters later took the logical step and became wholesale brewers as well.

However, this was not as easy a move as it later appeared, since the publican brewer proved more difficult to shift from his dominant position at the bar in Wales than his counterparts in most of Britain. In 1840 there had been 27,125 pubs in the UK brewing their own beer; fifty years later this was down to 6,350. This was a dramatic drop.

But in the wide licensing area of south-west Wales, administered from Carmarthen, official statistics in 1872 show that nearly two-thirds of the pubs in the region still brewed their own beer

– 731 out of 1,158 houses. And these pubs accounted for the vast bulk of the beer produced, using 82,473 bushels of malt compared to 12,528 handled by five common brewers.

In the Cardiff region where large-scale brewing was much more established – 54 common brewers accounting for 365,764 bushels of malt – still 1,214 pubs brewed their own beer in 1872 out of a total of 2,783 houses. More remarkably, they were using almost as much malt, 302,329 bushels.

The publican brewer lingered longer in rural areas and the old market towns. George Borrow came across many during his famous wanderings through *Wild Wales* in 1854, and usually found their beer to his taste, like the time he asked for a jug of home-brewed in a pub in Bala.

LLANFAIR CAEREINION,

8 MILES FROM WELSHPOOL STATION.

Wynnstay Arms Hotel,

FAMILY & COMMERCIAL HOUSE.

WINES & SPIRITS OF THE BEST QUALITY.

HOME BREWED ALES.

Families and Tourists will find every accommodation, combined with moderate charges, at the above Hotel.

GOOD RIVER FISHING IN THE NEIGHBOURHOOD.

EDWIN HUGHES, Proprietor.

ADDED ATTRACTION: pubs always thought it worth including in their adverts that they provided home-brewed ales

> I tasted it and then took a copious draught. The ale was indeed admirable, equal to the best I had ever before drunk – rich and mellow with scarcely any smack of the hop in it, and though so pale and delicate to the eye nearly as strong as brandy. I commended it highly.

Christopher Cobbe-Webbe recalled the bustling heart of Haverfordwest in Pembrokeshire on market day in his book *Haverfordwest And Its Story*, published in 1882:

> Our progress . . . is constantly interrupted by a horse-bench outside of every public house. These latter are very numerous, and from their doors there streams forth an everlasting smell of new drink, for the inmates are constantly brewing.
>
> A publican being asked concerning the age of the ale he was supplying to his customers, is said to have replied that it would be a fortnight old the following Thursday week. This, of course, was a joke, kept up against one poor man. Still it was literally from hand to mouth.

The author concluded: 'I fancy little but home-brewed is consumed.'

It was never easy for a publican-brewer to keep his customers satisfied, as Cobbe-Webbe's tale of a travelling judge testified. His Honour had stopped for refreshment at a well-known half-way house between Cardigan and Haverfordwest, the New Inn.

Here a flagon of cwrw dda was called for by his Lordship, who having tasted it with the air of a connoisseur, required the attendance of the landlord. Cap in hand, mine host appeared.

'Pray landlord,' said his Lordship, 'tell me where you procured the malt from with which you brewed this ale?'

'From Ha'rfordwest, my Lord.'

'Then where DO you get the water from?'

'From handy-by, my Lord,' said the gratified landlord.

'Ah,' said his Lordship, 'it is just as I thought. If you had to go to Haverfordwest for the water, and had the malt been handy-by, the ale, I opine, would have been much better.'

Haverfordwest was packed with pubs. There were fourteen alone in Dew Street at one time. Trade directories regularly stated that brewing was 'prosperously pursued' in the town, yet few local wholesale brewers emerged to challenge the publicans.

Matthew Whittow developed the Spring Gardens Brewery in the 1850s and 1860s – his fourpence a quart beer was said to have been very popular – but after Edmond and Rees had taken over the business around 1870 the firm stopped brewing, the new owners taking their ales from Allsopps of Burton-on-Trent. Alfred Beynon, William Jones and George Green also tried their luck in Bridge Street, but Green seemed to be hedging his bets. He was in addition a dealer in corn, agricultural seeds, clover, turnip, mangold and artificial manures. None survived for long.

In 1895 Kelly's Directory was still recording ten pubs that brewed their own beer: the Bristol Trader, The Quay; Farmer's Arms, Holloway; Fishguard Arms, Old Bridge; Fox & Hounds, Hill Street; Horse & Groom, Prendergast; New Inn, Old Bridge; Oak Inn, St Thomas's Green; Royal Oak, Dew Street; Swan Hotel, Swan Square and the Three Crowns, Hill Street. This was out of a total of 103 houses.

Not all brewing houses, of course, were licensed. A law report of 1897 reveals that there was a persistent amount of illicit brewing:

On Sunday, 7th March, the police searched a house at Cwmdegwell Street, Dogmells, occupied by Elizabeth Thomas and her daughter, after seeing several people entering and leaving the premises with beer.

In the back room they found two nine-gallon casks of beer on tap; in the back yard a complete brewing plant was discovered and in another room about a hundredweight of hops. Beer measures and other indications of an extensive trade were also found. Subsequently, at the Kemes Petty Sessions held at Eglwyswrw, before a full bench of magistrates, both women were charged with selling beer without a licence.

A previous conviction having been proved, on which occasion the brewing utensils were ordered to be confiscated, the defenders were now fined £25 and costs.

To the north of Pembrokeshire the dominance of the publican brewer was much more marked. In an area in and around Cardigan, including St Dogmaels, Cilgerran and

Aberporth, there were only eleven pubs NOT brewing their own beer in 1910. The other thirty were happily fermenting away. A common brewer never appeared in the town, though Cardigan maltster Samuel Young of Bronwydd House in St Mary Street must have been doing good business. By 1926 the number of home-brew houses had halved to fifteen.

Even around the main brewing centres like Wrexham, pubs continued to brew their own beer. The parish of Ruabon had eight out of sixty-eight houses in 1886. Slater's Directory of 1895 includes the King's Head and White Horse in Rhos, the Fox & Hounds at Newbridge and the Red Lion at Bangor-on-Dee. In Wrexham itself pubs like the Black Lion in Hope Street, the Hop Pole in Yorke Street and the Old Swan in Abbot Street were brewing in the 1880s, despite the large number of common brewers in the town.

CHALLENGING THE GIANTS: the Old Swan in Wrexham brewed its own beer despite being surrounded by much larger commercial breweries

Many more continued without comment; brewing was just seen as a declining part of running a pub. When the last official statistics were produced in 1914 there were 1,477 publican brewers in England and Wales. Between the wars the number continued to dwindle, as commercial brewers either bought up the houses or persuaded the licensees that it was not worth the trouble of producing their own beer.

But a surprising number survived in Wales. It was not only in the far west that many publican brewers continued well into the twentieth century. In and around Crickhowell, near Abergavenny, fifteen out of twenty-nine pubs were brewing in 1910, including four in Talybont-on-Usk – The Star, Traveller's Rest, Usk and White Hart. Eleven were still mashing their own in 1926. In Abergavenny itself six pubs continued to brew in 1934 – the Griffin, King's Arms, Lamb, Red Hart, Victoria Inn and the King David, which also made its own cider.

Their ales were generally respected, not least because poor publican brewers soon went out of business or out of brewing. In the early days, of course, each home-brew house had to compete with their rival publican brewer down the road. David Jones recalled the situation in Llanblethian, near Cowbridge, in nineteenth-century Glamorgan:

Our host John o' the Picton [licensee of the General Picton] emulates his neighbour Shon-y-Gwaith [John the Weaver, landlord of the King's Head] as to who shall have the best tap, and the competition for that honour causes each of them to add an extra bushel of malt to his brewing, to gratify the village topers.

A visitor to the large Island Green Brewery in Wrexham in 1911 commented:

The numbers of country inns which formerly brewed their own ale are decreasing, and the owners naturally look to Messrs Huntley and Mowat [owners of Island Green] who are able to supply them with an ale which is nearer to the much-admired Home-Brewed than most liquors upon the market.

A more telling sign of home-brew's reputation was the fact that many of Wales's leading commercial brewers thought it worth their while to produce a bottled beer called Home-Brew. Imitation is always the sincerest form of flattery.

The restrictions of the Second World War seem to have killed off most of the remaining home-brew pubs. One of the last in Wales was the Queen's at Cefn Mawr, which was bought by Border Brewery in 1944. Certainly none was left by the 1960s, though one of the four surviving houses in Britain, the Three Tuns at Bishop's Castle, was right on the Welsh border.

Oddly, the steam from farmhouse brewing stayed longer in the air, some in remote areas like the Gwaun valley in Pembrokeshire continuing to brew their own for special occasions to this day.

MASS PRODUCED HOME-BREWS: such was the standing of home-brewed ale in Wales that many commercial breweries produced a bottled ale under that name

COUNTRY BEER

One step up the brewhouse ladder were the country and market town breweries, many of which evolved from home-brew pubs or malting and milling businesses. They tended to grow gradually, unlike the later brewers set up to meet the urgent needs of rapidly developing industrial towns.

Delafield's Brewery in Abergavenny is a typical example. The King's Arms in Nevill Street is one of the ancient Monmouthshire town's many pubs, dating from the sixteenth century, standing next to the medieval wall by Tudor Gate. Inside are old oak beams and partitioning; outside a large painted plaster coat of arms of Charles II.

The pub would have brewed from its earliest days and from the beginning of the nineteenth century the innkeepers extended the business by malting behind the pub as well. It must have been a cramped operation. Some of the 'honey-combed' ceramic tiles used in the drying kiln can be seen in Abergavenny Museum. John Whitmore Blashfield, the landlord from 1822–35, was listed as a maltster, as was Mrs Mary Ann Brock, licensee in the late 1850s and early 1860s.

ROLLING OUT THE BARREL: Delafield's Brewery in Abergavenny in 1903, growing out of the old King's Arms in Nevill Street (Abergavenny Museum)

In 1862 the King's Arms was bought by Thomas Delafield and he soon extended the business again, becoming a wholesale brewer by 1864. By the 1870s Delafield was also trading as a wine and spirit merchant. The family business spread further. Daniel Delafield became a beer retailer in Union Road, while Thomas Alfred Delafield ran the Monmouthshire House pub in Ross Road from 1884. Later Daniel took over the The Sun pub in Cross Street. By the beginning of the twentieth century the proud company boasted three horse-drawn carts for delivering the beer. The pub appeared to have become primarily a brewery, the main window filled with bottles. The firm was especially known for their Nourishing Brown Stout in earthenware 'ginger beer-style' bottles.

During the First World War this small family empire was taken over by Claude Atkin – and quickly crumbled. The handful of pubs were leased to Wintle's Brewery of Mitcheldean, Gloucestershire, in 1924, and the King's Arms reverted to being just a home-brew pub under a succession of licensees including William Rowlands (1926) and David William Fenner (1934). It was later taken over by the Cheltenham and Hereford Breweries and today the pub still trades, but without the brewery.

Another Abergavenny concern that grew a little larger was Facey & Son. Samuel Henry Facey was a saddler from Brecon who moved to the White Swan in Cross Street, Abergavenny, in 1864, and opened a wine and spirit business. He soon extended into brewing, having a short-lived partnership with Frank Morgan of the Brecon Road Brewery, before setting up on his own as the Cross Street Brewery by 1875. He transferred his wine and spirit business to Market Street and in 1892 erected a new brewery there. A profile of the time said:

The present premises have been specially built for the trade. The main building contains a well-appointed suite of offices, sample-rooms, show-rooms and sale-rooms. At the rear is a compact brewery fitted with plant and machinery on the tower principle. The latter has been laid down by an eminent London firm, and is much admired by all interested in brewing operations. The capacity is that of six quarters and a force of ten practical men, under experienced super-intendence, is kept constantly employed.

The brewhouse was not large but well designed. Two maltings were run in Lion Street and Mill Street. The company also operated the only

NEAT AND COMPACT: Facey's small but well-designed brewery in Market Street, Abergavenny, around 1900 (Abergavenny Museum)

bonded warehouse in Abergavenny in the cellars of the market buildings. A sign of the firm's standing is that they were the official bottlers and agents for Bass and Guinness in the area. Their own draught beers, according to a 1903 advert, were XK Mild at 10d. a gallon and XX Mild at 1s.

The business steadily passed through three generations. After Samuel Facey died in 1904 his son Frank, a partner since 1892, took over. Following Frank's death in 1933, his son Frank Edward succeeded – and worked hard to expand the company.

A feature in the *Abergavenny Chronicle* in 1947 said that the last ten years had seen much development, with new fermenting vessels, cask-washing machinery and a modern bottling line installed. Four more pubs had been bought and the company's influence 'spread far beyond Abergavenny'. Three motor lorries delivered as far afield as Cardiff, Swansea and Tenby. For a local brewery this was a huge trading area. Production had been increased from a tiny 600 barrels a year (less than 12 barrels a week) to 3,000.

Then, as for so many family firms, came the moment of truth. When Frank Edward Facey died in 1949, aged only forty-eight, there were no interested relatives waiting in the wings. The business, with twelve pubs, was sold to David Roberts Brewery of Aberystwyth in 1950. Brewing immediately ceased though the premises were used for another ten years as a bottling store and depot.

Such small companies emerged and then disappeared all over the country, usually being swallowed by a larger brewery. The Castle Hill Brewery in Ewloe, near Hawarden, in North Wales, followed a similar pattern. Founded in 1844 by John Fox and his brother-in-law James Heyes in a building behind the family farmhouse, the brewery used springs in nearby Wepre Wood to obtain water. This very hard source gave Fox's beers their distinctive taste. Until 1905 coal for the brewery was also obtained from a shaft in the wood.

From 1846 until 1933 the brewery malted its barley at Swndwr kiln near Northop, which was leased from the Bankes family of Soughton Hall. Malting, as was traditional, took place during the winter months, using local barley mixed with cheaper Californian and Chilean varieties.

James Heyes died in 1870 and John Fox in 1886, the business being carried on by Fox's eldest son, also called John Fox. Later he was assisted by his sons, John Holmes Fox and Cecil Hunt Fox.

The family company soon built up an estate of fifteen to twenty houses, though pubs were frequently lost following pressure from the licensing authorities, who were eager to reduce the number of poor premises. Rather than incur the heavy costs of bringing them up to scratch, the Fox family accepted compensation. Their main trading area was Deeside and around Hawarden, Buckley and Mold.

About twelve to fifteen men were employed at the brewery in the early 1900s. They produced two regular beers – mild (later called special) and IPA – plus unusually for Wales, a strong Christmas ale called Lamb's Wool. The brewery also bottled Bass and Guinness, the outside brews accounting for the bulk of their bottling work. Their own beers were mainly sold on draught, selling in 1922 at 5d. a pint for mild and 7d. for IPA.

In 1947 John Holmes Fox died and a year later Cecil Fox sold the Castle Hill Brewery to Burtonwood Brewery of Warrington, with sixteen pubs. He preferred to concentrate on the

MAN-POWER: there was no fast-moving bottling line at Fox's Brewery at Ewloe in the 1930s just steady hand-operated machinery (Clwyd Record Office)

farm, where a herd of pedigree Guernsey cows was established. Brewing at Ewloe ceased and Cecil, a keen organic farmer, grew mushrooms in the brewery cellars.

A more striking brewery in the same county of Flint was St Winefrid's Brewery at Holywell, which was sited right on top of a legendary spring. The spout of water was said to have appeared on the spot where the head of St Winefrid rolled after being sliced off by a sword in the seventh century.

Slater's Directory of 1858 explained the fascination of the small market town:

> From an early date Holywell has been celebrated for its fine spring of water called St Winefrid's Well . . . It bursts forth with great rapidity from under a hill and rises into a basin twelve feet by seven . . . The roof over the well is finely carved in stone with the legend of St Winefrid . . . It is also hung around with crutches said to have been left by persons who were cured by resorting to the water.
>
> One circumstance asserted of this spring, which to some may seem incredible, will at any time be demonstrated to the visitor. The basin will contain about

240 tuns of water which, when emptied, replenishes itself again in two minutes. The experiment was tried for a wager on the twelfth of July, 1831, in the presence of several gentlemen of great respectability when, to their surprise, the well filled in less than the time mentioned, proving that St Winefrid's spring throws up more than 100 tuns of water a minute.

Porter's Directory of 1886 added: 'This spring . . . has never been known to freeze and forms a river of very considerable dimensions, which in its course to the sea works wholly or in part 11 large mills (for making Welsh flannels and tweeds) besides affording an inexhaustible supply of pure water to the inhabitants of Holywell and adjacent towns.' It was obviously an ideal spot for a legendary brewery.

As early as 1844 the town boasted five breweries. Gradually one came to dominate. Developed by Phillip Dykins, the St Winefrid's Brewery was run by John Lloyd Price from the mid-1870s. The premises, right by the well in Greenfield Street, also included a corn mill and a large waterwheel.

The company built up a tied estate of some forty pubs, trading on the fame of the holy well. They were not the only ones to exploit the reputation of the water. Owen Brothers established a neighbouring St Winefrid's aerated water factory, claiming in 1886: 'The waters used in the manufacture of these drinks are procured direct from the famous St Winefrid's Wells, which are admitted by scientists to be equal or superior to any other spring in the Principality.' A Holywell chemist, John Carman, even boasted that his 'world-famed' hair restorer 'will restore grey hair to its original colour and beauty'.

After John Lloyd Price's death in 1923, the trustees decided to sell off the forty-four pubs, the bulk going in 1928 in a lot of thirty-four houses for £30,000. Brewing ceased and the premises were sold the next year. Among the lots in 1929 were a further four pubs – and five delicensed houses owned by the brewery.

The problem of losing pub licences was not solely because of pressure from local licensing magistrates. Many of the country houses must have been barely viable. Or at best part-time businesses for the landlord, usually run by his wife during the day while he worked at another job. When the nearby Cambrian Brewery at Bagillt, three miles from Holywell, was put up for sale as a going concern in 1893, the sales particulars are revealing about the state of many of these country concerns.

Originally established by William Pierce in the late 1840s or early 1850s, the Cambrian Brewery took great pride in its 'celebrated Welsh ales', even having them analysed to prove their purity. One Sheridan Muspratt, MD, stated in 1864:

I have made a careful analytical examination of the jar of ale which your agent in Liverpool submitted to me for the purpose, and which was described as follows – 'Sample of Mr Pierce's 6d Ale, obtained from R Wilkinson, Preeson's Row, November 1, 1864.' I find in it only the products of malt and hops, that is to say it is unadulterated, and in strength and flavour it bears favourable comparison with the ales in its class which I have examined.

Despite that recommendation the mild Sixpenny Beer was not a massive seller. It was the only beer brewed at Bagillt and about sixty barrels a week were produced, only half the

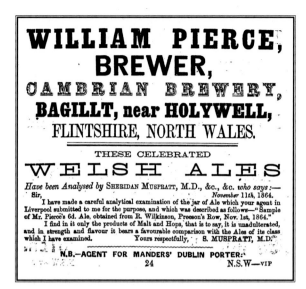

twelve-quarter plant's capacity. Tenants of the forty-one pubs (nearly all within six miles of the brewery) were free to look elsewhere for their bitter and stout.

Annual sales ranged from 150 barrels at the Castle Inn, Flint, to a tiny 16 barrels at the Hare & Hounds, Halkyn. The average was 50–60 barrels a year. Even if this is doubled to allow for sales of (more expensive) bitter and stout, this means most of the houses were selling little more than two barrels of beer a week. This was small business and hardly likely to encourage the brewers to invest much in their houses.

The state that some brewers' estates could slump into is sharply illustrated by Stiles Brewery of Bridgend.

Robert Henry Stiles had not enjoyed the best of fortunes after taking over the Glamorgan market town's only brewery around 1870. It had been built into an old woollen mill alongside the River Ogmore and in 1877 the river in full force flooded the town. There was six feet of water in the Wyndham Hotel, 'and a total loss of 300 barrels of beer from the brewery', reported the *Western Mail*. Worse was to come.

After the death of Stiles around 1880, the family company seems to have drifted. In the early part of the twentieth century Margaret Rosina Stiles was in charge. Some time in the 1920s the brewery stopped production, instead taking beer for their houses from Rogers of Bristol. In 1935 Simonds of Reading took over Rogers and discovered the bad news. Stiles owed Rogers over £15,000. Simonds' worried chief accountant went to investigate – and was appalled by what he discovered.

The twenty-seven pubs, he informed the Simonds' board, were 'in such a deplorably bad state' they would need large sums spent to improve them. An official survey of the properties valued them at £41,451. Some of the comments were scathing: 'Poor property – trifling trade.' 'Roof in very bad state.' 'Cheaper to rebuild.' 'White elephant.' Oddly the licensed vaults at the tumbledown brewery, though described as being in a 'wretched state – comprises large bare shop', were doing a 'good trade'.

The accountant hesitated to recommend buying such 'derelict properties' in satisfaction of the debt. He thought perhaps it was best to let things continue 'in the hope of the issue being forced by some other party'. Stiles also owed money elsewhere, but he did not hold out much hope. 'Any brewery company would be very chary about buying except at a knock-down price.'

Simonds only finally took the plunge in 1938 and bought Stiles. Though the houses were run down they did not want to lose the 3,000 barrels of business a year. To the east, in the heart of the Vale of Glamorgan, another country brewer survived a little longer.

Cowbridge has always been famous for its large number of pubs – there were twenty-three in 1851 for an adult male population of 274 – and many visitors have tried (and failed) to have a pint in each as they staggered down its long main road. Local folk said the High Street had a pub for every lamp-post. What kept the packed taverns and coaching inns going was the passing trade between Cardiff and Swansea and the thirsty crowds on market day. Iolo Morganwg even composed a verse in honour of the town's tipplers:

> For Cowbridge hath no sober man,
> Or none of milk sop thinkers,
> And no philosophical fools,
> But great and glorious drinkers.

With so many licensed houses (there were still twenty-one in 1914) a few commercial brewers inevitably appeared. Samuel Howells, one of a number of local maltsters, started brewing down Malt House Lane in the 1850s. John Evans ('Silurian' of the *Glamorgan Gazette*) used to visit the brewery to pick up some brewer's barm (yeast). As he later recalled, Sam Howells was a formidably striking figure.

Sam was one of the biggest men I had ever seen. He could easily hit a hole through the pine end of a house with his fist. He had been a well-known pugilist in his day and had large cauliflower ears and as many scars on his face as the map of England.

No one is ever known to have complained about his beer.

Another maltster, Lewis Jenkins, quickly followed Howells' example with the Vale of Glamorgan Brewery in the High Street. His son, licensee of the Butchers' Arms, started the Cowbridge Brewery in the 1890s further down the road by the bridge. This larger building was taken over by Thomas Morgan early in the twentieth century.

Morgan was a wine and spirit merchant and beer dealer in nearby Pontyclun (or Pontyclown as he spelt it on his beer labels, which sported a comical fat dragon standing on one leg). During the 1890s he also manufactured mineral waters in Pontyclun. By 1906 he had extended this liquid business into brewing his own beer at Cowbridge.

The venture was a success. Thomas Morgan & Sons Ltd, as the business became, bought pubs as far as Maesteg, Porthcawl and Bridgend, where they eventually owned six houses including the Dunraven Hotel. But the business was not without a note of tragedy. As Brian James and David Francis recorded in *Cowbridge and Llanblethian, Past and Present*, one ill-fated challenge in 1919 was not forgotten.

At this time Dillwyn Morgan of the Bridge Brewery, Cowbridge, owned a fine trotting pony called Daisy, whose coat gleamed like satin and whose speed was a byword in the borough. She was a dray horse who delivered barrels of ale from the brewery to every corner of the Vale. Mr Morgan was very proud of Daisy and boasted that his pony could outrace any horse in the locality.

He was challenged by a Llysworney farmer who owned a fast hunter. A private race was arranged at Penllyn, wagers were laid and an enthusiastic crowd of supporters turned up at the racecourse to cheer Daisy on. After a thrilling race she flew past the tapes ahead of her rival.

Such a voltage of triumph surged through the jockey's body that he hadn't noticed that Daisy had begun to tremble like an aspen. Within seconds, she collapsed on the racecourse and died. The remorse of her owner was so profound that he ordered an inscribed tombstone to be placed over her in the brewery yard – a lasting testimony to her gallant 28 years.

In peak times the brewery near Cowbridge Town Hall turned out 100 barrels a week, mostly of mild. Only in the 1950s, with increased competition from the larger brewers and production down to twenty barrels a week, did the company decide to sell up to brewing giants Bass.

The last brew was on 18 May 1955, supervised by Gwyn Thomas, head brewer for thirty-six years, helped by his two assistants Glyn Corbett and Jenkin Reid. The scene could have been from 1905, so little had altered. But outside, the brewing industry was changing rapidly.

The first country brewers in Wales to raise their vision beyond their local pubs came from Llangollen. Whether it was the excellence of the local water or the enterprise of one individual is unclear. What is certain is that by the mid-nineteenth century, this popular country town near Wrexham was known far and wide for its beer.

George Borrow, during his famous walk round *Wild Wales* in 1854, smacked his lips at the thought of Llangollen, 'which is celebrated for its ale all over Wales'. Slater's Directory of 1858 declared that 'Llangollen ale has attained celebrity all over the kingdom.'

The prime mover behind this achievement seems to have been Walter Booth's Llangollen Brewery, which by the 1850s was selling its beer far beyond its local boundaries. An advert of 1858 proclaimed:

The Llangollen Ale is already so well known and appreciated throughout England, that it is needless to add any further comment, and as they have stores in Manchester, Liverpool, Chester and Llangollen, parties in those or any other districts can be immediately supplied with any quantity.

This was a remarkable trading area for a Welsh country brewer. At the time the small town boasted another seven breweries, like Robert Baker's Sun Brewery, Attwell & Garrett's Victoria Brewery and John Davies's Red Lion Brewery. But most were small concerns, little more than glorified home-brew pubs, trading off the fame of Booth's enterprise.

The standing of Booth's Llangollen Brewery is shown by the fact that in 1870 it was taken over by a leading figure in the drinks industry, John Samuel Tanqueray, who came to live at Penybryn Hall. He had been a partner in the London brewers Combes (later to form part of Watney, Combe, Reid) and was a member of the distilling family, Tanquerays, still known today for their Gordon's Gin. He immediately ran into trouble when 'a destructive fire' severely damaged the premises in April 1870.

The small town was soon overshadowed by its larger neighbour, Wrexham, but its

END OF AN ERA: brewer Gwyn Thomas looks on as Jenkin Reid stirs the final brew at Thomas Morgan and Sons' Brewery in Cowbridge on 18 May 1955

PEACE OF THE COUNTRY: one of Facey's houses, the Red Lion at Llanbedr, after the takeover by John Roberts of Aberystwyth in 1950

reputation lingered on. *Heywood's Guide to Llangollen* of 1906 still claimed that: 'The beer is highly extolled for its many excellent qualities and is produced here in large quantities for the supply of Llangollen and the neighbourhood and for foreign demand.'

Tanqueray's Llangollen Brewery in Berwyn Road, near the A5 London–Holyhead road, closed just after the First World War, the pubs being bought by Soames of Wrexham. The premises were eventually taken over by the *Caldecot Press*. Robert Baker's Sun Brewery in Regent Street lasted a little longer, closing in 1925.

DEMON DRINK

T he Welsh brewing industry was different from neighbouring England. Besides arriving late – most of the leading commercial breweries did not appear until well into the nineteenth century – and then growing rapidly to match the demands of the rocketing towns in South Wales, it was shaped by another formidable force – the temperance movement.

In England the campaigners against alcohol held rallies, addressed meetings and argued strongly for restrictive legislation. In Wales they lived next door. Nowhere else in Britain did their beliefs take such a profound hold. Drinking almost became a guilty, back-door habit frowned on by all levels of society. Pubs – and even breweries – buckled under the pressure and closed. Those that survived tended to keep their heads down behind the bar. No nation boasted less about their beer than the Welsh.

This spirit sprang from the Methodist movement and the religious revivals of the nineteenth century. Their puritanical leaders preached self-sacrifice and abstinence from all worldly pleasures. Demon drink was top of the sin list. Temperance – moderate drinking – was not enough. Many demanded total abstinence from alcohol. The chapel and the pub, often standing uneasily side by side in the same street, became sworn enemies, competing for the same flock of followers.

The passion that could be provoked is shown by one writer, commenting on the closure of a brewery at Sennybridge:

> There was a large distillery [sic] and considerable trading was done in these damning commodities . . . It is a great blessing that this trade has disappeared and we hope that the day is not too far distant when the business will be but history in all our towns and villages.

Yet the church had not historically been hostile to beer. The first large-scale brewers had been the monasteries and many church ceremonies involved a drink or two. A nineteenth-century profile of Caerwys in North Wales recorded: 'There is in the possession of the vicar of the Church of St Michael an old zinc jug with date inscribed 1702 which had been used in former times to distribute ale at funerals.' Mourners traditionally drank liberally.

In medieval times church ales used to be brewed, their sale helping towards the upkeep of the premises. 'Ale at Whitsantide or a Whitsan Church Ale is a repairer of decayed country churches', said the poet John Taylor. Some clergy even ran pubs in the eighteenth century, depending on the income to help support themselves. The minister of Llanycrwys in Carmarthenshire kept an ale-house in the churchyard and was said to break off service to attend to customers on Sunday, according to Archbishop Tenison's visitation book.

UNEASY NEIGHBOURS: the chapel and the pub became competitors for the same flock in many Welsh towns and villages

Itinerant preachers were at one time rewarded with beer, a barrel sometimes being kept in the pulpit to lubricate their message. Vestry meetings considering church business were regularly held in pubs, with free beer given for unpaid services.

Some industrial areas developed so rapidly that in their early years, before churches were built, services were held in pubs. The long room at the Rhymney Inn (later the Beaufort Arms) in Rhymney was used by both the Church of England and Roman Catholics for Sunday worship by the first ironworkers.

There was no objection to beer in the early years of the Methodists, either. Henry Child, the leader of the movement in Llanelli and a close friend of John Wesley, founded a famous brewery there in the late eighteenth century. His son-in-law, the Reverend James Buckley, continued the business after 1824, giving his name to the brewery. The Three Cranes pub in Pontypool was run by a prominent Calvinistic Methodist elder. The gospel and the glass seemed to go happily together. But from the 1830s the mood began to change.

A key factor was legislation. In 1830 the Wellington government, anxious to reverse a growing spirits trade (and worried about a possible revolution), passed the Beer House Act allowing any householder, on payment of a two-guinea licence, to sell beer. Previously the number of licensed inns had been controlled by the authorities including the church. Suddenly it was open season. Almost anyone could run a pub. Within six months some 25,000 beer-houses had flung open their doors. The number of pubs in Britain nearly doubled in less than a decade.

At the same time the duty on beer was abolished. Thus ale was both cheaper and much more widely available. The growing gangs of miners and industrial workers, more affluent than their country cousins despite appalling working conditions, reached out eagerly for this fresh draught. Drunkenness spread, rolling and stumbling down the streets. The church was appalled. Temperance preachers found a swelling audience.

Another reason why the anti-drink bandwagon could now roll was the common cuppa. Tea had grown in importance in Britain from sales of one million lbs in 1730 to forty-six million by 1847. Here at last was an alternative drink safe from infection. Consumption of weak table ale fell by half in the fifty years to 1830. The British were losing the habit of enjoying small beer for breakfast.

The development much annoyed the famous brewery chronicler Alfred Barnard, who expounded at length on the qualities of beer following his visit to Buckley's Brewery in Llanelli in 1890:

> Ale is certainly the best and cheapest drink a working man can take. Tea, coffee and new milk are all higher priced than beer – that is if the non-intoxicating beverages mentioned are made of good quality.
>
> Excellent beer like that we tasted is especially suited for those hard workers, foundrymen, miners and engineers who abound in this populous district. It can be obtained at three pence the quart and forms a better article of drink with their solid food than tea.
>
> From our point of view, neither coffee nor tea can be allowed to stand in competition with good honestly brewed beer; nevertheless our object is not to denounce those useful beverages, but to commend ale as the heartier and cheaper for working people.

There is a desperate, defensive note to this tirade, and it is no coincidence that it was made in connection with a Welsh brewery, for it was in Wales that the pressure against beer was mounting the most. But not all brewers were alarmed at the sight of a teapot – and not all chapel-goers recoiled at the thought of entering a brewery.

The Welsh Wesleyan Methodist journal, *Yr Eurgrawn*, reported that after a fund-raising chapel meeting in Cardiff in 1846, Mrs Elizabeth Thomas, owner of the New Brewery in Great Frederick Street, invited the congregation round to her brewery at four o'clock for tea, 'which was very handy, and about 210 people partook of tobacco afterwards'.

The campaign against all 'strong drink' had first seen teetotal pledges introduced in 1832. Soon thousands were signing up, proudly hanging their framed certificates on their living-room walls. The nonconformist churches which came to dominate Wales took teetotalism to their heart. It was estimated in 1838 that in North Wales 600 ministers had signed the pledge. In five parishes around Bala in 1841 over 5,000 of the population of 6,000 were teetotallers.

John Jones of Rhyl wrote in one of the many temperance magazines in 1840:

None hardly enter the churches without being teetotallers. Yea, some churches refuse all but such, and have they not lawful ground to stand upon for doing so. Why? Because hundreds have been misled by these cursed drinks from their profession, and have backslided across the alcoholic half pints, and have been cast like Jonah, into the sea of intemperance, but teetotalism like the whale swallows them and casts them on dry ground; blessed be God for such a glorious cause. Hallelujah!

Such a fervent spirit not only dented beer sales, but also affected those working within the industry. At Aberystwyth eight publicans were numbered among one batch of 400 recruits. A brewer was said to have poured beer from his casks into the River Teifi. 'Some of the public houses are gasping for breath, as though they were in the last struggle', said a report from Abergwili in Carmarthenshire. Houses closed down including twelve in Bethesda and four in Morriston.

YOUNG CONVERTS: many children flocked to the temperance campaign – here demonstrating in Tonypandy in the Rhondda – but would their dry resolve continue in later years?

In the mining and iron town of Tredegar a New Temperance Hotel was erected 'principally by the working men of Tredegar, at a cost of nearly £3,000', stated Morris's Directory of 1862. It was first started by the great temperance advocate 'Cheap John', who visited the town in 1859 and 'induced the working population to abstain from intoxicating beverages'. It was opened on 2 December 1861. Such monuments to a new life without beer were repeated throughout the valleys – and worried the brewers.

During the temperance surge in Tredegar in 1859 some seven thousand signed the pledge. That summer the receipts of the Rhymney Brewery plunged by £500 a month. Such fervour did not last long. The *Star of Gwent* reported late in 1861: 'When the first intoxication of teetotalism was over . . . the working class, after most seriously alarming the publicans, returned to their cwrw and degraded habits.' But neither did the fervour disappear. It kept coming back, as wave after wave of religious and temperance revivals swept Wales.

The Rhymney Brewery, which had been shaken by the Tredegar experience, had been established by the directors of the local ironworks, with an eye to extra profit. But many industrialists were implacably opposed to beer as they believed that drunkenness seriously affected their business.

The Dowlais Iron Company at Merthyr Tydfil tried to employ an all-teetotal workforce. Ironmaster John Guest certainly did not want any of his men involved in the licensed trade. He ruled in 1831 that 'no person employed in our service must have anything to do with keeping a public house or beer shop, and if any person now does so he must be warned to leave one of the two.' He also tried to close pubs around his Dowlais works. He was particularly pleased in 1852 when he succeeded in shutting the Walnut by the gates 'which was considered the greatest nuisance'. Throughout the 1860s the company employed solicitors to oppose applications for new pubs in the area.

In Cardiff the movement went much further in bricks and mortar terms than just erecting a hall. Between 1850 and 1864 a large 'Temperance Town' – street after street without pubs – was built on land reclaimed from the River Taff, starting only a bottle throw from the front of Brain's Brewery in St Mary Street. The town's citadel, Wood Street Temperance Hall (later the Congregational church), was able to seat more than three thousand.

A host of organizations sprang up under the temperance flag. The 1907

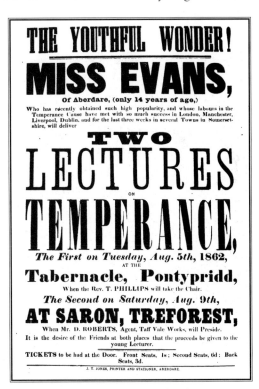

THE YOUTHFUL WONDER!

MISS EVANS,

Of Aberdare, (only 14 years of age,)

Who has recently obtained such high popularity, and whose labours in the Temperance Cause have met with so much success in London, Manchester, Liverpool, Dublin, and for the last three weeks in several Towns in Somersetshire, will deliver

TWO LECTURES

ON

TEMPERANCE,

The First on *Tuesday, Aug. 5th, 1862,*

AT THE

Tabernacle, Pontypridd,

When the Rev. T. PHILLIPS will take the Chair.

The Second on *Saturday, Aug. 9th,*

AT SARON, TREFOREST,

When Mr. D. ROBERTS, Agent, Taff Vale Works, will Preside.

It is the desire of the Friends at both places that the proceeds be given to the young Lecturer.

TICKETS to be had at the Door. Front Seats, 1s; Second Seats, 6d; Back Seats, 3d.

J. T. JONES, PRINTER AND STATIONER, ABERDARE.

TEACHING HER ELDERS: the young not only joined temperance organizations – they also lectured older folk on their bad habits

Cardiff trade directory listed eight including the Good Templars, the Rechabites, the South Wales Auxiliary, United Kingdom Alliance, National Commercial Temperance League, the Sons of Temperance and the British Women's Temperance Association.

The crusaders regarded themselves as a righteous army fighting against a dark evil. Some like the Blue Ribbon Army and the Salvation Army took their military trappings so seriously they literally fought for their cause, scrapping in the streets against a drinkers' opposition dubbed the Skeleton Army.

Harry Davies in his book *Looking Around Llanelli* recalls the militant Band of Hope marches in the town at the turn of the century. 'It was a demo calculated to make an impact on the conscience of the most hardened topers.' Led by the town's bands in full blast, they carried a wide range of banners – Pleidiwch Sobrwydd Dros y Tir (Vote for Sobriety throughout the Land), Unwch Fyddin Dirwest (Join the Army of Temperance) and Dwr Yw Ein Diod Ni (Water is Our Drink).

For a determined believer nothing was spared to defeat the enemy. Even to the extent of burning down a brewery.

The Mona Brewery at Llanfachraeth on the Isle of Anglesey had been established by John Jones by 1841. Twenty years later it appeared to be a modestly thriving business, with fifteen people living on the premises. It was of sufficient size to attract comment from Slater's Directory of 1868: 'Llanfachraeth is a small village. The only object worthy of notice is a large brewery in the occupancy of Mr John Jones.'

He died in 1870, his widow Margaret taking over the business helped by their son William. An advert of 1874 for Jones & Co. of the Mona Brewery said they were maltsters and wholesale brewers of ale and porter. Customers were warned that 'Llanfachraeth ales are so noted for purity and flavour that purchasers should take care they receive the genuine article when they apply for it.'

After the death of Margaret Jones in 1882 the Mona Brewery Company Ltd was formed to run the business. Registered with a capital of £5,000, its first share subscribers were mainly local farmers plus, surprisingly, 'Rev G.J. Hughes, Llanenchymeod, clerk in holy orders'. He also became one of the first directors. The venture does not appear to have been a success and by the end of the decade David Williams was the proprietor.

Events of much more significance were happening outside the brewery. The premises were part of the Garreglwyd estate and in 1890 the estate owner Sir Charles Reade died at the early age of thirty-eight. He was said to have been an alcoholic and to have drunk himself to death.

His widow Maria (Lady Reade), twelve years his senior, became violently opposed to the sale and manufacture of alcohol. In the following years she attempted to close all the pubs on the extensive estate and built a coffee house at Llanfaethlu to encourage more sober habits. Her campaign shook the Mona Brewery and put paid to plans by David Williams to sell the brewery. Even a successful case in Chancery could not force a sale.

It emerged in court that in 1889 David Williams had come to an agreement to sell the brewery to a local resident, Captain Nicholas Spargo, for £1,800. Spargo had been intending to float a public company, but following Sir Charles Reade's death he attempted to withdraw from the deal. Williams, desperate to sell, took him to court in 1892. The judge ruled in Williams' favour but Spargo never seems to have taken over.

Meanwhile, Lady Reade was determined to ensure that Llanfachraeth never brewed again.

There were still sixteen years of the Mona Brewery's lease left to run in 1892. She didn't mess about with the courts. According to local legend she torched the brewhouse. 'Whether there is any truth in this I do not know', says building researcher George Lees. 'However, there is some evidence of a fire in the main part of the brewery, now almost demolished.' Certainly the extensive premises never brewed again, being converted into a farm.

After her death in 1917 when the Garreglwyd estate was broken up and sold, a clause was discovered written into the deeds of the Mona Brewery buildings that they could never be used for the manufacture or sale of intoxicating liquors. This clause is still in force.

The Mona Brewery was an extreme case, but most Welsh breweries trimmed their sales in the face of the strong temperance tide. Frederic Soames of the Nag's Head Brewery in Wrexham (later Border Breweries) made a revealing comment when visited by brewery chronicler Alfred Barnard. He said that when he took over the business in 1879 he 'determined to brew a light thoroughly finished beer . . . because such ale was bound to assist the cause of real temperance'.

Brewing a temperance beer sounds a contradiction in terms, but what he meant was that he was seeking to produce a weaker, lower-alcohol ale. The fledgling Wrexham Lager Company took this one step further. They claimed that their brew was 'almost non-intoxicating'. Adding: 'When more generally known and consumed, it will diminish intoxication and do more for the temperance cause than all the efforts of the total abstainers.'

Wrexham was a famous brewing town – the Burton of Wales – known in its early days for its strong Welsh ales. 'The strongest variety of Welsh ale (and a marvellous production it is) is the Royal Wrexham', reported Barnard on his visit to Soames' brewery. This rich barley wine survived to become one of Border's best-known bottled beers. But by then it was almost the only genuinely strong ale produced in Wales.

Most Welsh breweries came to concentrate on producing low-gravity beers. In some areas the preference was for light-coloured brews – usually known as PAs – and in others for dark ales, but their common characteristic was that they were relatively weak. When they needed a premium ale, many breweries tended to buy one in from one of the large Burton companies like Bass.

Leading Welsh firms like Brains of Cardiff were out of the ordinary in producing their own premium draught beers as well as lower-gravity brews. Barnard thought it sufficiently unusual, to be worth comment: 'Messrs Brain & Co. have now become so proficient in the art of brewing pale ales that the Cardiff publicans are not obliged to go to the "beer city" for their Burton ales, that commodity being now brewed on the same principle in their own town.' Even then, Brain's best-selling beer for most of its history was the low-gravity Red Dragon dark.

David Williams' Taff Vale Brewery in Merthyr Tydfil also broke out of this weak beer barrel, commenting in 1893:

Some five years ago nothing else but fresh beers were practically consumed, namely cheap XX; but within the last three years the firm have been working up a large quantity of better quality beers to meet the present demands of the public, who now prefer a higher quality beer, such as pale and high-class bitters. The firm's leading productions are pale ale, golden hop bitters, old Welsh ale, XXXX old, XXX, XX, AK Stout and OK.

Soft Option

The temperance campaigners faced a major problem. If miners and other workers were to be persuaded not to drink beer — then what would they slake their thirst with? Various botanical and temperance brewers were established to try and solve this problem. Hancock's for a while sold a non-alcoholic beer known as Kops Ale. Howell Davies of Abercynon produced non-intoxicating Quaker Pep with a rugged miner on the label. But as Rhymney Brewery realized few workers were impressed with these brews. In 1922 they commissioned famous cartoonist H.M. Bateman to produce a cruel cartoon for their calendar, showing unhappy miners in Pussyfoot's Saloon trying such delicacies as 'Pansy Extract for Miners'. Rhymney Brewery's advice was: 'Stick to Buchan's.'

With this new range the company won medals for their beers in 1890, 1922 and 1923.

But they were the exception; most Welsh breweries relied for the bulk of their sales on low-gravity brews. In part this was because many of the brewers were serving miners or iron and steelworkers, who needed to drink in quantity to satisfy their raging thirsts. But similar areas in Britain, like the Durham coalfields and the steel industry of Teesside in north-east England, produced stronger brews. The steelworkers of Hartlepool sweated out of the furnaces to sup pint after pint of Strongarm from local brewers Camerons. As its name implies it was far from a weak beer.

The difference in Wales was the strength of the temperance movement. Breweries felt on the defensive and did not want to be accused of producing heavy, intoxicating ales. Caution was catching.

The leading beers all declined in strength over the years, often heavily weakened by government restrictions during the wars. But unlike in other areas, many basic brews failed to bounce back at all after restrictions were lifted. Hancock's PA had a gravity of 1051 in 1894; 1046 in 1905; 1041 in 1925 and 1040 in 1935. After the Second World War the Cardiff-brewed beer was down to a gravity of 1034 in 1945 and 1032 ten years later.

Border Breweries of Wrexham became brewers of a variety of milds. Even their Border Bitter (1035) was of below-average strength for a bitter beer. In South Wales the story was the same. Fernvale Brewery of Pontygwaith in the Rhondda was brewing just one draught beer in 1960, a light SPA of 1031 gravity. When a stronger (1036) Cambrian Bitter was introduced in 1965 it only sold in a few outlets.

Henry Holder, one of the last head brewers at Rhymney Brewery, recalled that they produced two draught beers at Rhymney – Golden Hopped Bitter (1036) and Pale Mild Ale at a very low gravity of 1030. The weak PMA was by far the best seller, averaging 1750 barrels a week compared to 250 for GHB.

Even when South Wales' brewers produced so-called 'strong ales' in a bottle, they were notoriously weak. The term had become so discredited by the 1920s that Buckleys of Llanelli advertised their bottled Strong Ale under the slogan 'The Strong Ale that IS Strong'. Even then it wasn't. It was just less weak than the others. Hancock's bottled Strong Ale in 1960 had a gravity of 1030.

Wales's taste for low-gravity beers has lasted to this day. Darker brews like Brain's Dark and Worthington Dark are still very popular. The country's two major brands are Allbright (1033) from Welsh Brewers and Welsh Bitter (1032) from Whitbread. Both processed beers reflect in style and strength the traditional light-coloured PAs which once dominated the market. A market heavily influenced by temperance pressure.

Keith Kissack in his book *Victorian Monmouth* gives an account of the growth of the temperance movement in one town, in an area which was then neither part of Wales nor so dominated by nonconformist principles.

The editor of the local Beacon newspaper even thought that the Total Abstinence Society, founded in the town in 1839 with 100 members, was absurd. 'Let the working man, whose bodily energies have been strained to the limit . . . be asked how he could exist without his daily beverage of beer, and he will tell it will be impossible, his strength will fail him.'

But even in the more relaxed atmosphere of this genteel county town, the temperature could quickly rise. A reformed drunkard denounced Queen Victoria before a crowd of 350 at the

Market Hall for not wearing a teetotal medal. 'That would have been a prouder ornament than the crown which rested on her brow. She would indeed be a queen worthy to reign who had reigned over her own passions and appetites.' Many were upset by this attack on the young queen and the next meeting was disrupted 'by the disgraceful employment of missiles'.

In 1877 the vicar of St Mary's Church formed a company to set up a pub with no beer for working men. Monmouth had over sixty licensed houses and topped the pub league table in Britain in 1902 with one house for every eighty-three people. The British average was 242. The vicar's venture bought the Reformer's Tavern to provide tea, coffee, food, dominoes, billiards and well-aired beds. The British Workman opened with a fanfare, but failed to intoxicate local workers and folded within two years.

The movement in Monmouth suffered a major setback in 1883 when at the burial of a former mayor, Mr Coates, the lengthy funeral procession was prevented from entering the church by the vicar, Reverend Wentworth Watson, on the grounds that spirits had been taken into Mr Coates's house during his illness. The vicar said this meant that Mr Coates was not entitled to full burial rites. He conducted a shortened service in the graveyard. This hardline attitude caused an uproar and alienated many. The St Mary's branch of the movement was disbanded.

Temperance campaigners were not above stooping to dirty tricks. When at the turn of the century there was a widely-publicized case of arsenic poisoning in beer in Manchester, Monmouth crusaders attempted to smear the nearby Redbrook Brewery. The company was forced to place a large advert in the *Beacon* proclaiming that their ale was 'absolutely free from arsenic'.

HOLY WAR: Revd Penry Thomas objected to beer posters on Cardiff trams in 1932, prompting this cartoon by J.C. Walker in the South Wales Echo. *Despite the ridicule the vicar had his way and all drinks advertising was banned*

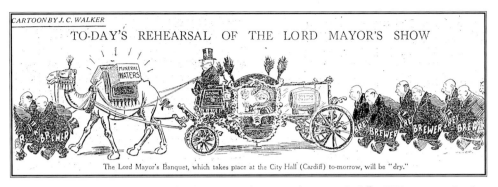

CARTOON BY J. C. WALKER

TO-DAY'S REHEARSAL OF THE LORD MAYOR'S SHOW

The Lord Mayor's Banquet, which takes place at the City Hall (Cardiff) to-morrow, will be "dry."

FUNERAL PROCESSION: the brewers pay their last respects to beer at civic functions in Cardiff in 1932, as portrayed in the South Wales Echo

The issue was still at the heart of local politics in Wales in the 1930s, causing heated exchanges over minor matters. In 1932 Cardiff's tramways committee decided to discontinue drinks advertising on their buses. Beer was even banned from the back of tickets. The city's Licensed Victuallers' Association objected in disgust. 'We are rather tired of the continued attempts to use the council chamber as the medium for furthering teetotal propaganda, much of which in the past has made Cardiff the laughing stock of the country.'

The year 1932 was to see an even bigger coup for the temperance movement in Cardiff when a teetotaller was elected Lord Mayor. Alderman C.F. Saunders declared that there would be no alcoholic drinks at receptions and functions during his term of office. He pledged that it would be 'a dry civic year'.

When the dry Lord Mayor's Banquet took place in November, the *South Wales Echo* cartoonist J.C. Walker depicted the brewers as mourners at a funeral, following the Lord Mayor's carriage decked out like a hearse in which was slung a dead barrel of beer. Soon opponents were claiming that the corporation had lost thousands of pounds since the drinks ban at the City Hall – and drinkers were sneaking in their own bottles anyway.

The strength of feeling in Wales against drink is best expressed through the novels produced by the Temperance Publishing Company. *The Tavern Across the Street* by Derwenydd Morgan, published in 1915, is a classic example. The author wrote 'to encourage the great army of temperance workers, who are today fighting this twentieth-century enemy – the curse of drink in Wales'. He came from Pencader, between Carmarthen and Lampeter, but set his book in an imaginary Welsh mining town called Cambria.

The tale revolved around the downfall of a jovial farmer, Sam Rosser, who came to Cambria to run a pub, the Fox & Geese. Soon the house was leaning against the gates of hell. 'The drinking, the swearing and the fighting on many a Saturday night in the Fox and Geese were enough to make humanity shudder and Christianity weep tears of blood: and I often wondered that the bolt that struck Sodom did not strike them, or that the earth would refuse to carry the burden any longer and would open and swallow them.' The landlord's daughter Matilda killed herself after being wooed and abused by the heavy-drinking young Squire Hancock.

The pub was represented as the source of all evil. As one character argued: 'When we go to the pub and spend money, what do we get? We get beer and whisky, in other words, we get bad headaches, shattered nervous systems, ruined constitutions, divorces, suicides, murders, empty cupboards, naked backs; homes without furniture, without love. These things and many more do we get by patronising the bar-room.' He added: 'I wish they had prohibition in Wales, the same as they have in some of the states of America, then there would be no temptation.'

The temperance workers, led by a reformed drinker, won victory in the town elections. The Fox & Geese declined.

> The people who called at the tavern now were those who had reached the lowest strata of society – the station known the world over as the gutter – the last stage of the drunkard, men who had nothing more to lose for their all had been sacrificed on the altar of strong drink – an ogling, drinking, thieving class, who ought to have died long ago when all that was manly in them died.

The pub shut and was turned into a temperance hotel.

A stranger visiting the new idyllic Cambria asked: 'How did you get the tavern out of the town?' The reply was, 'By getting the people out of the tavern, and the tavern out of the people.' But author Derwenydd Morgan was not happy to leave his novel there. He was looking forward to an even brighter future.

The final chapter was devoted to a meeting called to organize a Welsh Prohibition Party 'to endeavour to prevent the manufacture and sale of the accursed liquor which is the cause of all this poverty and woe'. His hero Alwyn Pryse believed: 'This traffic is like a devastating prairie fire that must be extinguished once and for all: like an overwhelming deluge from which the people must be rescued – like unto a plague whose ravages must be stayed.' He was convinced that 'every clergyman, every minister of the gospel, every Christian worthy of the name will respond to the bugle call of Prohibition.'

The novel concluded:

> It will be a strenuous fight and a bitter one, but victory will be ours. Ere many years roll by the black flag of rum will come down and every tavern will be closed. The white flag of Prohibition will wave on every hill and in every dale, and our children growing up in Temperance Wales will wonder that we tolerated such an institution as an open tavern in our land.

It was against such a fanatical background that breweries in Wales plied their trade. It was never as comfortable a business as in England.

NEVER ON A SUNDAY

With the temperance movement's solid support from most of the churches and chapels, it was no surprise that the campaign soon came to focus on the sensitive question of drinking on Sunday.

It was an issue that had long concerned those in authority. With pubs and churches uncomfortably close in many towns and villages, community leaders did not like walking in their Sunday best to divine service past noisy beer shops, with staggering drinkers spilling out over the pavement. Drunkenness was bad; heavy drinking on the Lord's Day was tantamount to blasphemy.

The government took an early interest. Previously pubs had been allowed to open when they wished, though some areas had introduced their own limited local rules. This freedom began to disappear on Sunday. The Alehouse Act of 1828 decreed that licensed premises should not trade during the hours of divine service. This vague requirement was replaced in London in 1839 by the Metropolitan Police Act, which prevented pubs opening before 1 p.m. on Sunday. This more easily controlled law soon spread to other areas and was then extended to the whole country in the Lord's Day Act of 1848.

In Calvinist Scotland this was not enough. In 1853 the Forbes-Mackenzie Act introduced complete Sunday closing in Scotland, the only exception being hotel and innkeepers supplying lodgers and genuine travellers. This 'bona fide' traveller requirement was later to haunt Wales.

Soon temperance and Christian leaders in England and Wales were wanting more peace on the Sabbath than just the morning. An Act of 1854 introduced afternoon closing from 2.30 p.m. until 6 p.m. with evening closing at 10 p.m. Drinkers, who had not objected too strenuously to their longer lie-in on Sunday mornings, began to revolt.

After rioting and demonstrations in Hyde Park in 1855, an amended Act was hurriedly introduced increasing the lunchtime session from an hour-and-a-half to two hours, with a new closing time of 3 p.m. The evening period was extended by two hours, with opening allowed from 5 p.m. to 11 p.m.

The year 1869 saw the end of the free-for-all in setting up pubs. The Wine and Beer House Act gave licensing justices the power to control all licensed premises and refuse new licences. The ever-growing number of beer-houses – they had been increasing by 2,000 a year – was brought under control. Between 1830 and 1870 more than 53,000 had been established. Now their numbers began to decline.

Licensing legislation was a dangerous arena even for heavyweight politicians. Prime Minister Gladstone, who increased penalties for liquor offences in his Licensing Act of 1872, blamed his election defeat by Disraeli in 1874 on the brewers. He told his brother he had been 'borne down in a torrent of gin and beer'.

In 1878 came the Irish Sunday Closing Act which, while shutting all houses in the countryside, took pity on thirsty city dwellers. Bars in Dublin, Belfast, Cork, Limerick and Waterford were allowed to open from 2–7 p.m.

Welsh temperance campaigners were determined not to lag behind Scotland and Ireland, but were delayed by the reluctant next-door neighbour. Wales was lumped in with England as far as legislation was concerned. It was not until 1881 that the Principality broke free and obtained its own Sunday Closing Act.

The success at Westminster was hard won. The vast majority of Welsh MPs, especially the Liberals, strongly supported temperance reform, but there was little history of legislating separately for Wales. The victory for the anti-drink forces was based on considerable support within the country, which eventually proved impossible to deny, even if it took a trick in the end.

Agitation in favour of Sunday closing had been growing throughout Wales during the late 1860s and early 1870s. A Swansea conference in 1875 passed a resolution 'that a measure be introduced into Parliament specially for Wales providing for the entire closing of public houses on the Lord's Day'. Spearheaded by the Good Templars, the bandwagon began to roll. Petitions proliferated.

ON THE MARCH: women and children take to the streets in Wales in support of temperance legislation in the House of Commons

A huge canvass of nineteen Welsh towns during 1878–80 claimed that 38,443 households were in favour of the proposed measure, with only 717 against. In Aberdare it was claimed that 92 per cent of mining families were in favour. Such figures were certainly selective and probably exaggerated, but there was no escaping the fact that in combining respect for the Sabbath with restrictions on drinking, the temperance movement in Wales had found a winning combination. Public support was overwhelming. The only noticeable opposition surfaced in the cosmopolitan town of Cardiff, where the Licensed Victuallers' Association gathered 16,844 signatures against.

The Bill was introduced into Parliament in 1879 by John Roberts, MP for Flint. He declared that 'the continuance of the Sunday traffic is impossible without doing violence to the Welsh nation generally'. Significantly, the main opposition in the Commons came from English MPs worried about the Bill's creeping effect. Delaying tactics and the pressure of other business meant it was blocked in 1880. But Roberts's private member's Bill was introduced again in 1881 and this time nothing, it seemed, could stop it.

Gladstone decisively argued that Wales should be considered as 'a distinct case' on this issue, and the second reading was passed with a majority of 146, only 17 voting against. Edward Carbutt (Monmouthshire) unsuccessfully tried to get his county included in the measure. Hardinge Giffarde tried to exempt his constituency of Cardiff to no avail. Then the Bill ran into a parliamentary brick wall.

The dominant measure of the day was the Irish Land Bill. There appeared to be no more time for the Welsh Bill. It missed its third reading on 6 July. The brewing industry breathed again – and then howled in fury. On Saturday, 20 August, the Bill was slipped through near midnight by Gladstone when only seventy MPs were present. Opponents bitterly protested about this 'trick of legislating by surprise'.

The brewing industry regarded the Act with fear, as 'prohibition by degrees'. It hit them hard. Sunday was said to be the pub's most profitable day. Workers had more money in their pocket and time on their hands then, than on any other day of the week.

Some observers felt there was much more to the Act than just restricting drinking. The *Western Mail* believed it was the first instalment of home rule – a hope nurtured by many supporters.

W.R. Lambert in his book *Drink and Sobriety in Victorian Wales* stated that 'the Welsh Sunday Closing Act of 1881 was the most remarkable legislative achievement of the Welsh temperance movement in the nineteenth century'. He could have added – or of any other century. It was to prove the peak of their progress.

Welsh MPs could have pressed home the advantage by supporting a similar measure for England. Bills were repeatedly introduced throughout the 1880s, but when the crunch came they held back. In 1889 twenty-one Welsh MPs abstained on the second reading of the English Sunday Closing Bill, which was defeated by seven votes.

The operation of the Welsh Act generated tremendous controversy during its first decade – not least because on its success or failure was seen to rest the fate of England. It first came into operation in the Caerwys and Caergwrle petty sessional divisions of Flintshire on 27 August 1882. Immediately miners from Caergwrle walked to the Hawarden division where the pubs were not closed until 10 September. In South Wales early implementers were Llandaff and St Nicholas in Glamorgan, Narberth in Pembrokeshire and the country areas of Carmarthen. Services of thanksgiving were held in many churches.

J.C. Griffith-Jones, writing in the *South Wales Echo* over fifty years later in 1933, cast a damning judgement on the Sunday Closing Act:

> For years before . . . the sparks of controversy flew fast and furious. For years after the Bill was passed the fireworks continued.
>
> Instead of checking the drink evil which was its avowed intention it had the remarkable effect of increasing drunkenness and letting loose some ugly elements among what had hitherto been a sober and orderly people. A new era, the Shebeen Era, was ushered in.

He wrote of the 'disorder and devilment' of the 1880s and 1890s. 'Illicit drinking became a grand urge', as liquor was sold in private houses and underground drinking clubs – the shebeens. In summer Sunday 'field clubs' multiplied. A stream of hikers staggered over the border from dry South Wales into wet Monmouthshire.

> The scenes were indescribable. When time was called on Sunday nights the highway between Newport and Cardiff was turned into a Bedlam. A vast army, more merry than bright, rent the night with ribald songs and rolled all the way home. Many slept off the effects of these orgies in ditches or awoke to a hopeless dawn in a police cell.
>
> In Cardiff there were fierce Sunday night clashes between the rebels and the police in Mary Ann Street, Gough Street and Adelaide Street. Pokers as well as invective were in evidence.

Yet this wild impression of drunkenness and lawlessness was not reflected in the personal testimonies of leading citizens given in a book published in 1899, *The Case for Sunday Closing*, compiled by the Sunday Closing Campaign. Former mayor, Alderman Trounce, recalled notorious Bute Street in Cardiff docks before the passing of the Act:

> This street was at that time on Sundays too often a reproach to civilisation, the scene of drunken brawls and rowdyism. With public houses at every corner, frequented by sailors and others of all nationalities, respectable pedestrians were frequently subjected to gross insults and the sound of most offensive language.
>
> I have visited Bute Street several times since the Act in question became law, but the scene has changed – order and quiet prevail where once there was so much disorder and riot. Our police have easy times as compared with their onerous and trying duties of 20 years ago.

Another former mayor, Alderman Sanders, said that Cardiff 'was like heaven upon earth on Sundays since the passing of the Act'. Lewis Williams, JP was equally adamant: 'As senior licensing magistrate of the borough, I have no hesitation in stating that it has been an untold boon in promoting the good order and morality of the town, especially so in the lowest parts.' He concluded: 'Where lawlessness prevailed we have now almost perfect order.'

Both opposing views were wide of the mark. Finding objective reporting of the working of the Act was almost impossible, so partisan were the two sides. But reading between the lines of propaganda and avoiding the flying spears of suspect statistics, a hazy picture emerges.

Club de Marl

The most amazing and open act of defiance against the Sunday Closing Act was the Club or Hotel de Marl. Workers used to gather on a piece of waste ground – the Marl – in Grangetown, Cardiff, where rubble had been tipped after a dock extension. Small barrels of beer would appear and everyone relaxed over a few drinks. A sheet was spread on the ground and each visitor cast his sixpence or bob into the kitty of this open-air gathering. There was singing but no serious rowdiness. The police looked on but generally did not interfere, much to the annoyance of temperance campaigners. Eventually, the owner of the land, Lord Windsor, was persuaded to block access and the Club de Marl folded.

In the countryside and the market towns, where the Act was widely supported, its provisions were largely upheld, though there was undoubtedly some 'back-door' pub drinking. In one community knowing locals asked each other, 'Are you using your pencil today?' You needed one to lift the latch on the closed pub door. But in the larger towns and mining regions, where drunkenness had caused more alarm, evasion of the law was widespread. This took two forms, depending on the attitudes of local magistrates.

In some areas drinkers took to the roads. Genuine travellers were allowed refreshment on Sunday under the Act. To become a 'bona fide' traveller you needed to cover a distance of at least three miles. So suddenly pubs on the edges of towns and out in the immediate countryside began to do a roaring trade on Sunday. Thirsty folk, for instance, moved from Swansea to the Mumbles.

However, these leg-weary customers were supposed to be there on necessary business. Magistrates and the police could take action if they believed they were only out for a drink. People living close to the border could always, of course, pop over to Monmouthshire (which did not adopt the Act until 1921) or England.

More worrying for the police was the amount of drinking driven underground into illegal drinking clubs, the shebeens. Small casks of beer – in some places mockingly called John Roberts's after the MP who introduced the Sunday Bill – were bought as late as Saturday night and set up in private unlicensed houses. The beer was often adulterated and highly priced, but no one dared complain.

The streets might be quieter, but behind closed doors the law was being ignored and falling into disrepute. Like Prohibition in America, the Sunday Closing Act fostered a criminal culture widely supported by many members of the public. There were said to be 500 shebeens in Cardiff alone (including over 130 in Temperance Town!).

Many official clubs also sprang up, whose members had little in common apart from a thirst on Sunday. Between June 1882 and June 1883 more than three thousand workmen joined clubs in Cardiff, an increase of 900 per cent on the previous year.

W.R. Lambert concluded in *Drink and Sobriety in Victorian Wales*:

> The Act was successful in rural areas and in the towns of North Wales, where it was least needed, but met with a great amount of evasion in the urban communities of South Wales. The inconvenience and hardship of a closed public house on a Sunday was acutely felt in parts of Wales, as beer, the national drink, was not as portable nor as easily preserved fresh as was, for example, whisky, the national drink of Scotland and Ireland, where Sunday closing was also in operation. Consequently, the demand for drink on a Sunday manifested itself more clearly in Wales than in Scotland and Ireland.

The Cardiff morning newspaper, the *Western Mail*, led the attack on the Act, claiming in 1888 that Sunday drunkenness had increased. Lord Aberdare, who had guided the Bill through the House of Lords, challenged the editor Lascelles Carr to substantiate this claim. Carr responded by instructing his assistant editor David Davies to carry out a survey. The findings, not surprisingly, depicted a boozy Sabbath seen through the bottom of a beer glass. What was surprising was that Lord Aberdare swallowed this view.

> The effect . . . of the picture painted by your correspondent is such that were a Bill introduced into Parliament for the repeal of the Welsh Sunday Closing Act, I could not, unless evidence were adduced of better effects produced by a better administration of the law, vote against it.

The admission was a bombshell – and one which the *Western Mail* immediately seized on. The paper's proprietor J.M. Maclean, Conservative MP for Oldham, demanded an inquiry on 19 March 1889. The Home Secretary agreed to set up a Royal Commission to look into the working of the Act. The Welsh temperance movement was furious, particularly since they felt that the 'packed' commission was biased against them. They believed the five-man panel under Lord Balfour, including four Tories, would 'sell a nation to please a party'.

Both sides immediately began to prepare their evidence. Many brewers came forward. County Councillor Samuel Brain of the Cardiff Old Brewery blamed the Act for the rise of the shebeen when he gave evidence in October 1889. The Act 'certainly drove the people into illicit drinking', he said. 'In this town it really is impossible to imagine the men will not get beer, even if they have to get it in an illicit way.' He described how the system operated:

> Since the Act came into operation in 1882 nearly all the brewers within a radius of 100 miles of Cardiff have sent local agents here, and local men have had to go in very largely for the small cask trade. The great majority of the private houses in certain districts here have in immense quantities of this beer on Saturday nights, and in fact right up to one or two o'clock in the morning.
>
> It was nothing unusual for the well-conducted licensed victualler to stand at his door on Saturday nights, after he has closed his premises at 11 o'clock, and see brewers' drays and agents' carts delivering 70 or 80 small casks into private houses within 100 yards of his premises.

He believed these illegal dens would be 'stamped out' if pubs were allowed to open reasonable hours on Sunday.

Mr Lloyd, a traveller for William Hancock in Cardiff, demonstrated the extent of this trade. He gave figures for the company's three breweries in the town (the Bute Dock, Phoenix and South Wales). These showed that their sales of firkins and pins (nine and four-and-a-half gallon casks) had risen from 872 casks in 1880–1 to 39,055 in 1888–9. It was a massive jump. Nearly all this business was conducted through a new breed of street-corner wholesaler. Hancock's own pubs took much larger barrels (36 gallons) and hogsheads (54 gallons).

Alfred Watkins, a brewer in Hereford, confirmed that distant brewers were exploiting this new market. Selling through local agents, small casks now accounted for 69 per cent of his business in the Principality. In 1881 they formed no part of his trade in Wales. In Cardiff alone the figure was even higher at 92 per cent.

Watkins also said that his business in bottled beer had greatly increased.

SUNDAY SUSTAINER: the two-pint flagon bottle became a distinctive feature of breweries in Wales, as drinkers demanded something large enough to see them through the dry Sabbath

Many households were stocking up to see them through the dry Sabbath. Welsh breweries responded by producing their own special container – the flagon or two-pint bottle.

While the Royal Commission was taking its evidence in Cardiff in October 1889, some two thousand workers marched through the town demanding 'British Rights for British Working Men' and 'No Compulsory Sunday Closing'. In the following month an election in the West Ward was dominated by the issue. The Tory E.J. Smith, campaigning for 'The Liberty of the Working Man', was returned with a substantial majority. He was the president of Cardiff Licensed Victuallers' Association.

In North Wales Frederic Soames of the Nag's Head Brewery in Wrexham told the commission of a different development – an increased trade in his own houses owing to 'the abuse of the bona fide travellers clause'. He said: 'The Act in its objective of increasing temperance has most distinctly failed.' He presented a petition signed by over six thousand workers calling for it to be repealed.

The temperance movement also lobbied hard, with many petitions and well-briefed witnesses. They won the argument hands down, despite their earlier fears about the bias of the panel. The commission's report was all they could have wished for – and more. It concluded:

One of the suggestions most frequently made to us was the repeal of the Sunday Closing Act, or its modification in the direction of repeal by permitting the opening of public houses for a short time in the middle and evening of Sunday, either for sale both on and off the premises, or for the latter only. We cannot, after giving them the fullest and most careful consideration, endorse either of these recommendations.

Had it been our duty to advise on the form of the original legislation we might have suggested that some facilities should have been given for obtaining drink in small quantities for domestic consumption. We are, however, convinced that a change in this direction would be so unwelcome to so vast a majority of the population in so large an area of the Principality, that we do not think it ought to be forced on this large area for the sake of a possible benefit to the rest of the country. Moreover, we find an almost complete absence of evidence of a desire for such an amendment on the part of those classes who would be most likely to require or use it.

In other words, outside of Cardiff, some other parts of Glamorgan and around Wrexham, Wales supported Sunday closing. Most of the land was firmly behind the temperance movement. It was a sobering thought for breweries in the country. Their livelihood was at stake.

The Royal Commission recommended tightening up the Act to remove abuses. Police powers against shebeens should be increased; clubs more closely controlled. Travellers on Sunday must prove they were out on legitimate business and licensees had to record the name and address of all such customers for police inspection, or lose their licence.

The temperance campaigners were riding high and had already moved to achieve their next objective before the commission appeared. In 1886 a meeting at Rhyl had drafted a Local Option Bill for Wales which would allow areas to vote for complete prohibition of alcohol or at least a reduction in the number of pubs.

Radicals like Lloyd George of Cymru Fydd (the Young Wales movement) saw the issue bound up with the question of Wales's nationality – why should moral, sober Wales be linked with debauched England? In 1891 he claimed that three-quarters of Welsh people were in favour of total prohibition.

The first Local Option Bill for Wales was introduced into the Commons in 1887. Despite many delays and disappointments, a third Bill saw its second reading carried by six votes in 1891. Lloyd George boasted about 'our splendid victory'. The *Alliance News* exulted: 'The outpost has fallen and the capture of the main citadel is but a question of time.' They crowed too soon.

The brewers and distillers could not allow this 'thin-end of the wedge' to succeed. Once local option was introduced into Wales it would spread to England. Prohibition would creep across the land and their beer trade would be poured away. The Bill went into committee on 19 March and the Marquis of Carmarthen, a director of Hollands Gin, moved a sweeping amendment. Held up and blocked at every stage, the Bill was finally dropped on 14 July.

Another Bill was introduced in 1893 and met the same fate in committee after gaining its second reading with an improved majority of thirty-six. One supporter, Herbert Lewis of Flintshire, claimed 'every cubic inch of the House of Commons is charged with obstruction'.

It was to prove the last real chance of success, as Welsh politics became absorbed by the disestablishment of the Welsh church, putting the problems of the parson before the publican. Temperance legislation was elbowed aside. By 1894 Lloyd George was saying that home rule must come first and then prohibition in Wales would follow once the country had escaped 'the brewers' ring which seems to govern England'.

Though the temperance movement split and faded after the mid-1890s, Lloyd George was later to gain his revenge on the brewers under the guise of emergency measures during the First World War, when he was Minister of Munitions. His famous quote – 'Drink is doing us more damage in the war than all the German submarines put together' – sprang from years of frustration.

Under the Defence of the Realm Act (DORA) of 1915, a new Liquor Control Board was allowed to impose strict licensing hours for every day of the week in areas it deemed militarily important; in some places like Carlisle breweries were taken over and closed down, the pubs being placed under state control. In Wales, Cardiff, Newport and Barry felt the brunt of the new regulations.

From 17 August 1915, opening time within the three towns and for five to eight miles around, was restricted to 5½ hours a day. Pubs could only sell beer from 12–2.30 p.m. at lunchtime and from 6–9 p.m. in the evening on weekdays. On Saturday they had to put the towels over the pumps an hour earlier at 8 p.m. The period allowed for off-sales was reduced still further. Around a thousand pubs were affected.

For customers used to being served from 5.30 in the early morning until late at night, this was a crushing restriction. Dockers threatened to strike at Barry in protest. Harry Prickett of the Cardiff LVA said the order spelt 'absolute ruin' for many licensees. The brewers generally kept silent, not wanting to appear unpatriotic. But W.H. Brain, president of the South Wales Brewers' Association, described the orders as 'a great blow to the trade'. A more outspoken colleague, George Westlake, blamed 'fanatical teetotallers' who were using the war to push their 'fiendish propaganda for the purpose of wiping out the trade'. As if to

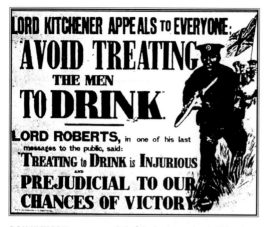

LORD KITCHENER APPEALS TO EVERYONE:

'AVOID TREATING

THE MEN

TO DRINK'.

LORD ROBERTS, in one of his last messages to the public, said:

TREATING to DRINK is INJURIOUS and PREJUDICIAL TO OUR CHANCES OF VICTORY

DRY VICTORY: temperance beliefs had taken such a hold by the First World War that even buying a drink for a friend was regarded as dangerous

add insult to injury, petty rules were added. One banned treating – in other words no one was allowed to buy his friend a drink, or even lend him money for one. Rounds were strictly prohibited.

Anxious about possible compensation, the brewers at first asked their houses to stay open their usual hours even though they could not serve alcoholic drinks. It was not a success. One pub reported selling only a packet of cigarettes in the morning. The house had a bar staff of five. One licensee gave an account of his new start to the day on the first morning when the regulations came into force:

The doors of the hotel were opened at 5.30 a.m. with a 'business as usual' spirit . . . The bar has been cleaned and polished overnight and the mullers, bright and shining, are filled and ready with good, piping-hot tea and coffee. In full view of any customer that may enter is a display of Bovril and biscuits. All is ready to supply the demand for anything in the refreshment business excepting intoxicants.

The hotel clock shows that it requires just five minutes to six when one of the sons of toil enters and asks for a rum. The request is met with the answer, 'Sorry, sir, but you cannot have rum. Will a cup of coffee or tea suit?' 'Coffee or tea! Strewth! I want something to stay my stomach with.' 'Well, how would a glass of hot milk do?' 'Look here, boss, I finished being weaned long ago and, besides, I have no time to stay over long hot drinks – must get on the job.'

After several more absent-minded and disappointed callers – but no sales – the landlord retires for breakfast and leaves the empty bar to his staff. 'Upon returning, I find that the quietness is only disturbed by the ticking of the clock. Next I read the takings by the cash register tills, but find all with the figures at zero.' He examines the sawdust on the floor for footprints but, 'like Robinson Crusoe of old,' discovers that 'I still reside on a lonely island.'

Later in the day came a call. 'A stone ginger, please.' Upon receiving the drink the customer remarks that he thinks he will pinch his nose whilst drinking, 'so that I cannot taste the damn stuff'.

The ban on treating also took some getting used to, as the *Western Mail* reported:

The words 'What is yours, old chap?' come almost unconsciously to many lips, and the warning look shot across from the other side of the bar counters invariably ended in a good laugh and a solo order. 'Botheration to it, I say', said one barmaid, who had to make ten times the usual number of journeys to the cash register. 'It makes my arm ache giving out so much small change and serving one customer at a time . . . No, Mr Jones (turning to a regular customer), you cannot give Mr Jenkins a drink. He must go on his own'.

If trade was heavily down in Barry, Cardiff and Newport, the regulations meant extra business elsewhere. Pubs just outside the restricted area heaved with new customers. Magor near Newport was flooded, the *South Wales News* reported in September, 1915:

When I visited the place yesterday (writes one of our representatives) the little village was crowded with brakes, traps and cycles. In addition a large number of persons had travelled by train . . . When I reached the railway station I realised that the description of 'Magor like a fair' was mild, for the scenes baffled description. Men and women were lying about on seats and platform; others were imbibing freely from bottles which they had brought with them. *Tipperary* and *Genevieve* and other popular airs rent the air, while a few engaged in fisticuffs.

The regulations soon covered much of the country and after the war the restrictions were only relaxed to a limited extent. The 1921 Licensing Act allowed pubs to open for eight hours a day, and Sunday closing was extended to Monmouthshire. At least buying a round was no longer illegal.

Ironically these closer controls probably saved British breweries from the fate of their colleagues in America where prohibition was introduced. In Britain the authorities came to believe that they had solved 'the drink problem'. Certainly widespread drunkenness staggered off the streets. In part this was because beer never regained its stupefying strength of pre-war days. In 1914 the average gravity was 1052; by 1919 it was down to 1039 and never recovered. Today the average gravity is around 1037. The Depression and higher prices caused by increased taxation also kept consumption down between the wars.

The Chief Constable of Cardiff reported that the number of cases for drunkenness in the city in 1932 was 158, down 56 on the previous year. In 1897 there had been a peak of 1,667 convictions. Newport reported their lowest number of cases on record (83). Merry Merthyr, with 178 pubs, had only 35. Swansea was almost off the legless table with just 17 convictions in 1931. Barry was down to 22 in 1932.

'It is gratifying that excessive drinking is unquestionably a relic of by-gone days', commented Mr J.C. Meggitt, chairman of the Barry licensing justices. 'A drunken person is seldom seen on the streets. The figures are incredible compared to pre-war years (when Barry had more than 1,500 convictions a year).'

This marked change brought an equally dramatic transformation of official attitudes. When the next war arrived in 1939 beer was no longer the enemy but an ally, vital for maintaining morale. Temperance demands for strict control were dismissed out of hand. Quintin Hogg, later Lord Hailsham, bluntly told the dry lobby:

The Temperance Council must clearly understand that the national emergency is not a moment to introduce temperance propaganda under the cloak of national necessity. Beer is the innocent pleasure of many millions, especially among those who bear the brunt today.

The pub was now seen as 'a block-house on the home front'.

The tide of opinion had turned completely against the temperance movement. Prohibition, discredited and dropped in America, was no longer on the political agenda. Pressure instead was turned on the campaign's main gain in Wales – Sunday closing. Even famous teetotallers came out against the dry Sabbath. Field Marshal Viscount Montgomery caused a stir in 1954 when he told a tourist development dinner at Cardiff Castle that Wales must end the ban on Sunday drinking. 'That is living in the past', he said. Many others felt the same.

The eighty-year drought came to an end in 1961 when a new Licensing Act allowed every area to decide their own policy by referendum. Immediately large parts of Wales voted to open pubs again on Sunday. Others followed as ballots were held every seven years,

RELIEF AT LAST: regulars at the Gwyn Hotel in Pontardulais watch as landlady Iris Hammett pulls their first Sunday lunchtime pints after the 1961 poll voted to open up the bars again in Glamorgan. But just 50 yards away, over the Carmarthenshire border, the pubs remained shut (Western Mail)

until by 1982 only two areas remained dry. In 1989 one of these, Ceredigion, switched to the wet camp, leaving Dwyfor on the Lleyn Peninsula out on a limb.

In 1993 Home Secretary Kenneth Clarke announced that he was scrapping the polls, the final one to be held in 1996. Last orders had been called on the once powerful temperance movement – which at times had come close to shutting down the Welsh brewing industry.

MASH TONS

arge-scale brewing depended on large towns or large-scale industry. Only when there were enough thirsty throats crowded together, did it make sense to pour investment into bigger breweries, particularly at a time when land transport was poor. Beer was bulky, heavy and awkward to move around.

In England, these sizable concerns first appeared in the greatest concentration of people – London. Indeed, the giant porter breweries of the capital like Whitbread and Truman were some of Britain's largest industrial enterprises of the eighteenth century.

In Wales, large-scale brewing initially sprang up alongside the ironworks and coal mines, most of them not arriving until well into the nineteenth century. But the country could have boasted its own eighteenth-century brewing giant – Arthur Guinness.

The famous stout brewer, who had founded his firm in 1759, had run into trading difficulties in Ireland. The powerful London porter brewers were shipping their beer into Dublin at a considerable advantage, undercutting their local rivals. The English Revenue Laws so penalized the Irish brewers that many went bust. Arthur Guinness thought it better to get out of Ireland rather than go out of business. He told a committee of the Irish House of Commons, set up to look into the crisis in 1773, that he intended to establish a brewery in Caernarfon or Holyhead – if he could find one ready-built there. Or he would build one if the unfair laws stayed as they were for the next few years. His aim would then have been to ship the beer back into Ireland, so avoiding the harsh taxes.

This may seem an empty threat since there were certainly no suitable premises at this time along the North Wales coast. But Arthur Guinness did actually visit the area and seriously considered settling there. If the regulations had remained in force he may have had no choice but to move his business across the Irish Sea. Sadly for Wales, the laws were relaxed in Ireland and Guinness's St James's Gate Brewery in Dublin went on to become the largest brewery in the world.

Wales did have one large-scale brewery in the eighteenth century, sited alongside its first major industry, copper smelting. Swansea was one of the cradles of the industrial revolution, Dr John Lane having moved there (because of its cheap coal) in 1716 to establish the Llangyvelach copperworks, shipping in copper ore from Cornwall. Soon other works followed turning the lower Swansea valley along the banks of the River Tawe into a smoking hive of activity years before the first blast furnace glowed in anger in Merthyr Tydfil or the pits were sunk in the Rhondda. Workers flocked to the area. Brewers followed, some coming from eastern England to exploit this growing new market.

In 1792 Swansea Corporation granted one of the valuable river bank sites to Messrs Phillips & Kendall to build 'a very considerable brewery'. This Cambrian Porter Brewery on the Strand became a prominent feature of the port and soon gained a rival. Reverend

Richard Warner in his *Second Walk through Wales* of 1799 talks about 'two expensive breweries' in Swansea.

The Cambrian Brewery passed through many hands: William Shepherd in 1798; Henry Bonham of London in 1801; Samuel Hawkins of London in 1802 and then to George Haynes in 1805. Five years later the lease was transferred to John and Francis Morse of the Oak Street Brewery in Norwich. But their interest seems to have remained in Norfolk, where the Morse family became widely involved in brewing, for by 1822 Haynes & Morgan were in charge.

According to Pigot's Directory of that year there were two breweries in Swansea both on the Strand: the Cambrian under Haynes & Morgan and another run by Edward David. Neither lasted beyond the mid-century as their sites in the docks became too valuable. The focus of the town's brewing trade moved away from the river, notably George Rolls developing what was later called the Swansea Old Brewery in the Mysydd Fields.

Henry Child built his brewery in the centre of neighbouring Llanelli in 1799 (later to become Buckley's Brewery) and that year also saw the first commercial brewery in Wrexham, the town which became the 'Burton of Wales', its streets dominated by breweries on almost every corner. The iron industry was prominent around Wrexham, the Wilkinsons building up the Bersham works from 1753, specializing in producing cannons. But these were early pioneers. The vast majority of Welsh brewing companies did not appear until the mid-nineteenth century or later, when the rapid development of the iron and coal industries, particularly in South Wales, created a huge new market for beer.

*RAILWAY PIONEER: a monument to Richard Trevithick's historic steam engine of 1804 in Merthyr, with the ruins of David Williams' Taff Vale Brewery on the skyline (*Western Mail*)*

This market started to appear as early as the 1750s with the first ironworks across the heads of the Glamorgan valleys. These works expanded swiftly after 1775 when James Watt's steam engine was made commercially available. In the 1780s arrived such influential figures as John Guest, who became a partner in the Dowlais works, and Richard Crawshay, who leased the Cyfarthfa works in 1786. But their development was held back by transport problems to what were, in those days, remote and inaccessible areas. Roads were poor and pack-horses and wagons inadequate for shifting huge loads of iron ore and coal, not to mention the finished iron products. The powerful ironmasters forced a major transformation. Between 1790 and 1820 a network of canals was constructed linking much of industrial South Wales to the coast. The most important, the Glamorganshire canal between Penydarren, near Merthyr Tydfil, and Cardiff, was opened in 1794. Its creation proved vital to the later development of Cardiff as a major port.

Horse-drawn tramroads connected some of the works to the canals and ran between the waterways, and it was along one of these stretches that Richard Trevithick's first locomotive steamed in 1804 between Penydarren and Quaker's Yard, heralding the start of the railway age. It was said to have drawn ten tons of bar iron at 5 m.p.h., but did not come into general use for carrying goods until ten years later.

Once technical difficulties were overcome, railway mania followed. The first line in Glamorgan, the Taff Vale Railway, was completed in 1841, bringing the great ironworks within an hour's travel of Cardiff, where the Bute West Dock had been opened for business in 1839. At the top of the line was Merthyr Tydfil, which had grown almost overnight into the largest town in Wales. In 1841 it had a population of 35,000, over three times the size of Cardiff. It was a forbidding sight.

'Merthyr Tydfil is situated in the highest part of Glamorganshire, among rough, bleak mountains. The neighbourhood abounds with heaps of cinders . . . the heat of which causes them to smoulder for long periods', wrote Kelly's Directory. 'The scene at night is beyond all description; the immense fires give a lurid hue to the faces of the workmen, and cause them to present a most ghastly appearance; while the noise of the blast and steam rolling mills, and of the massive hammers worked by steam machinery, preclude the possibility of being heard when speaking.' You can sense the sweat – and the raging thirst.

Pigot's Directory lists just one common brewer in 1822 and 1830, Williams & Bryant of Jackson's Bridge, who were also maltsters and millers. Most of the early demand for beer was met by publican brewers. Clearly they could not cope for long. By 1848 Hunt's introduction to the town stated: 'The breweries in the town and neighbourhood are upon a very extensive scale.' The Directory listed ten. Like the population, the brewing industry had grown at a rapid rate. By 1871 there were fourteen breweries for a town boasting 180 pubs.

One of the oldest was Giles & Harrap's Merthyr Brewery. This company had first surfaced at Tydfil Well under Watkin Davies in 1830, changing hands a number of times before John Giles took charge by 1852. Within twenty years it was trading as Giles & Harrap. The premises covered two acres with a frontage 250 ft long on the Brecon Road. Besides brewing milds, bitters and stouts, the company became famous for its extensive wines and spirits business with bonded stores in Bristol, London, Cork, Belfast and Greenock (but none in Wales). One of its leading brands was FOH (Fine Old Highland) whisky.

Its main rival was David Williams' Taff Vale Brewery, which had been founded by Thomas Evans as the Taff Brewery in 1843 between the canal and the river. It was bought by David Williams in 1867 and in 1888 he took into partnership his nephew William Griffiths. David Williams was prominent in the town, being three times in succession High Constable as well as on the Board of Health and the Board of Guardians. He founded the local Chamber of Trade before his death in 1897. The 25-quarter brewery was replaced in 1904 by a completely new brewhouse built on a prominent position above the town. The company boasted that its bonded stores in Merthyr were the only ones outside Cardiff in the area.

Just beyond Merthyr the Pontycapel Brewery at Cefncoed became so well known that it was one of only four breweries in Wales visited by Alfred Barnard while writing his famous work, *Noted Breweries of Great Britain and Ireland*. He found its setting quite idyllic on his visit in 1890:

> Leaving the environs of the town, where the operations of smelting iron ore have made the surface look like a vast range of volcanoes, we soon reached the open country, with its pleasant and diversified scenery. After passing through the village of Cefn, we descended a steep hill, from which our first view of the Pontycapel Brewery was obtained, and were much struck with its delightful situation and the scenery in the immediate neighbourhood where, on a small scale, all the charms of nature are united.

The brewery alongside the roaring River Taff was unusually powered by a large water-wheel. Its brewing water came from a famous spring known as Ffynon Oer or the cold well. Even more distinctive was its range of beers. Pontycapel specialized in stronger ales. One

ALE AND WELL MET: the Pontycapel Brewery at Cefncoed which charmed Alfred Barnard in 1890 (Welsh Folk Museum)

brew in particular probably drew Barnard to this distant location, as he revealed during his tour of the premises:

In the basement of the brewery, which we next visited, is situated the vathouse containing eight lofty stout vats, each upwards of 12 feet high and used for storing porter and stout, a beverage for which this company is so famous . . . Passing from thence into a passage sunk below the yard, we came to the old beer cellar, an arched chamber 55 feet long, containing four enormous vats for maturing old beer.

But he was seeking something else.

Next to the great vat room at the back of the brewery stands, by itself, a triangular building nearly 100 feet long. It is paved with stone, opens out onto a pathway by the side of the river, and contains ten lofty vats, one of which is called 'Good Friday', that name being painted on the front to distinguish it from the others, it having been fixed in its present position on that day. It is used for storing the celebrated Pontycapel XXXX ale, so largely in demand in the locality.

He had found what he was looking for.

The XXXX brew was known as 'Star Bright' – 'a beverage celebrated for its strength and flavour, not only throughout the South of Wales, but in the neighbouring counties of

GONE FOR A BURTON: the Rock Brewery in Aberdare which was taken over by Allsopps of Burton-on-Trent (Cynon Valley Libraries)

England'. Barnard was equally impressed with the IPA, 'as it is bright, nutritious, well flavoured with hops and quite equal to any we tasted in Burton'.

The Pontycapel Brewery had been founded by a maltster, Robert Millar, in the 1840s, but was developed by a retired Herefordshire merchant, Thomas Pearce, who bought the company around 1860. On Barnard's visit it was owned by his son, Harry Pearce, but run by a brewery manager, Thomas Morris.

To the west in the next valley, Aberdare also developed several breweries, notably the Trecynon, the Black Lion, the George and the Rock Brewery. But here, as elsewhere throughout Wales, there was a worrying development for the local concerns. Towns like Merthyr and Aberdare had expanded rapidly thanks to improved communications, particularly following the arrival of the railways. But improved communications worked two ways. While coal and iron could be shipped out much more easily, so other goods could just as easily move in. Among them was beer.

Burton-on-Trent had grown to become the beer capital of Britain thanks to its ideal water supply – and the railways. The companies there needed to sell the bulk of their beer outside the small Staffordshire town to survive. South Wales, with its mushrooming population, was an ideal market. Soon every railway arch around the main railway stations seemed to be packed with barrels from Bass, Allsopp, Ind Coope, Worthington, Salts and other Burton brewers.

Photographs of nineteenth-century Welsh pubs never seem complete without window signs for Bass or Worthington. Their excellent beers developed such a widespread following that

THIRSTY WORK: miners settle down for a well-earned pint in Cwmbach, beneath a mirror advertising Bass's Burton Ale

Dec: 6th, 1899.

ALTON COURT BREWERY Co.
LIMITED.
Ross. Herefordshire.

To Messrs Blythway & Son,

Solicitors,

Pontypool.

Brewers, Maltsters and Mineral Water Manufacturers.

Registered Telegraphic address, BREWERY, ROSS.

AGENCIES AT

BERGAVENNY	COVENTRY	CHELTENHAM	COLEFORD	GLOUCESTER	LEDBURY	LYDNEY	PETERCHURCH	SWANSEA	TEWKESBURY
BEWDLEY	CAMBERWELL	CHEPSTOW	EWIAS HAROLD	GREAT MALVERN	LEOMINSTER	MERTHYR	PONTYPRIDD	TALGARTH	WESTON super MARE
BRECON	LONDON	CRUMLIN	FRAMPTON-	HAY	LLANBISTER ROAD	MONMOUTH	PRESTEIGN	TENBURY	WEOBLEY
BRISTOL	CARDIFF	CINDERFORD	ON-SEVERN	HEREFORD	LLANWRTYD WELLS	NEWENT	RAGLAN	TINTERN	WORCESTER
BROMYARD	CARDIGAN	CIRENCESTER	GLASBURY	KINGTON	LYDBROOK	NEWPORT	RHAYADER	TENBY	WARWICK

Dear Sirs,

" Lewis -- A.C.B.C."

I am much obliged by receipt of yours of the 5th inst:, but after your letter of the 29th Nov: and my interview with Mr Lewis yesterday at Crumlin, I cannot quite understand it, I presume therefore you mean that we are to have the first I4 years at £8, with the option of a further 7 years Lease at £I2 per annum, if this is correct it will be quite satisfactory to us. I am sorry I cannot return the Lease for alteration, as I have already had it signed, and had intended altering the term and the rent to suit, but await your reply before doing so.

We are desirous of commencing building operations at once, and should be glad of a reply by return of post.

Yours faithfully,

Llewooler

WIDE NETWORK: the Alton Court Brewery's agencies stretched as far as Cardigan and Tenby from Ross-on-Wye

REVEALING WINDOWS: adverts for Worthington on draught and in bottle are displayed at the Welsh Harp Inn in Aberdare

many local brewers found it simpler to sell the popular Burton pale ales, while concentrating their own production on lower-gravity milds and other 'bread-and-butter' beers.

Aberdare was no exception. As early as 1856 the Freemason's Arms was advertising Bass. In 1867 the Clarence Inn was offering Ind Coope, Allsopp and Bass. By the 1890s the Rock Brewery was acting as an agent for Worthington while the Black Lion handled Allsopps. In Aberdare this process eventually went one step further. Allsopps bought the Black Lion Brewery and the Rock Brewery. By 1921 brewing had ceased in the town, apart from the small George Brewery.

The Burton breweries were a major force in Wales. Even large companies sold their beers, Hancock's being closely linked to Worthington for many years, while Brain's handled Bass. Guinness and other Irish brewers came to dominate the stout trade. A few London breweries like Whitbread and Truman were also active in the Principality, though they arrived later in the nineteenth century. Birmingham breweries led by Mitchells & Butlers and Ansells made inroads into central Wales and Newport, while Liverpool, Manchester and Warrington brewers were prominent in North Wales. The directors of Lees Brewery in Manchester liked to buy hotels along the main fishing rivers as they were keen anglers.

Perhaps the biggest surprise was the lack of adventure from the leading Bristol breweries like George's who never made a determined effort to capture the South Wales market. The main Bristol brewery to push over the border was the smaller Jacob Street Brewery of W.J. Rogers, whose AK ales were sold through branches in Newport, Cardiff and Swansea. The Alton Court Brewery of Ross-on-Wye had an even wider network in Wales, with agents in many smaller towns.

Competition was fierce for the Welsh breweries and few managed to reverse the trend and sell their beers across the border into England.

IRON BREWMASTERS

The smoke from the foundries of Merthyr had soon spread over the hills to the east to the next valley, the Rhymney Valley. There beside the ironworks sprang up a brewery which was to overshadow all its neighbours.

Industry had first come to the bleak northern end of the valley in 1800, when Upper Furnace was built at Rhymney Bridge. Richard Crawshay of the Cyfarthfa Works in Merthyr Tydfil soon took over this site for his son-in-law, Benjamin Hall.

Hall, who ran the Union Ironworks until 1820, branded his name into history following his later government position as Commissioner of Works. He was responsible for introducing the guarantee stamp – the Hallmark – on precious metals. His name also rings out every time his great bell – Big Ben – booms over the Houses of Parliament.

In 1825 the Marquess of Bute set up a rival ironworks opposite in typically elaborate and eccentric style. The design of the 40 ft high furnaces was based on the Egyptian ruins of Dandrya. In 1837 the two amalgamated to form the Rhymney Iron Company.

The Bute Ironworks had employed a Scotsman, Andrew Buchan, a carpenter by trade, to deepen and straighten the River Rhymney to prevent flooding. His handling of his team of labourers impressed the iron company directors. Buchan paid his men partly in goods through a shop he already ran at nearby Twyncarno, and when the ironworks decided to set up their own shop in Rhymney at the end of 1836 they appointed the shrewd Scot as manager.

The truck system – where employees were paid in tokens rather than money, which could only be cashed at the company shop – had been made illegal in 1831. The system was wide open to abuse as workers had no choice but to accept the prices and quality of goods on offer. Some firms used it simply as a way of making extra profit at their employees' expense. Others were happy to let their men run up debts knowing they would not then be free to leave their employment. A few had been known to pay in ale at their own pubs.

Though these truck or tommy shops were illegal, the law was easily evaded in the Welsh valleys where the ironmasters and colliery owners ruled like kings, often building their own villages for a largely immigrant workforce. The company directors were frequently the local magistrates. Some were Members of Parliament.

At remote Rhymney the iron company virtually built the town, not only establishing a shop and houses but also a farm, church and school. They went one step further. The minutes of 7 February 1838, record: 'The chairman recommended the propriety of building a brewhouse for the supply of beer to all persons employed in the works.' It was quickly resolved by the board: 'That the recommendation of the chairman be approved of and carried into effect forthwith.' Furnace work was thirsty work.

A month later a cheque for ten guineas was drawn payable 'to Mr Kemp for drawings of a brewery'. By January 1839 tenders had been accepted from Pontifex of London for supplying brewing vessels, price £2,255. The brewery engineers left their mark on the walls with a metal plate carrying their address 'Shoe Lane, London, 1839'.

It was not the only connection with the distant capital. The chairman of the Rhymney Iron Company was MP William Copeland, a Lord Mayor of London, and the first board meeting was held in Laurence Pountney Hill, London. What was needed was a resourceful and reliable man on the ground in Wales – Andrew Buchan. He was appointed brewery manager in addition to his duties at the company shop.

Within a few years of being built the substantial buildings were described as 'one of the largest breweries in South Wales' with the reputation of the beers soon spreading further afield than the local ironworks. But the name attached to its Rhymney ales in trade directories was not Andrew Buchan, but first James Penny in 1844 followed by John Tozer in 1848. Buchan was appointing others to run the brewery; he was more interested in the shop.

His premises were no street corner affair but a grand emporium. He might have a captive market – deductions were made direct from wages for goods supplied – but he gave value for money. People came from miles around to shop there, besides the local workers. The place was more like a department store. It sold everything a working man might need. You could buy moleskin trousers, woollen mufflers, cloth caps, hobnailed boots and iron-ringed clogs. There were pots and pans, enamelled mugs and plates, heavy iron kettles, oil lamps, scrubbed wooden tables and ironframed beds. Proud families wanting to outshine the neighbours could obtain Welsh dressers and grandfather clocks. And outside there was a whole industry meeting its more basic needs.

Weekly supplies of eggs, cheese and butter were brought in by carts from neighbouring farms, some owned by Buchan. Horse-drawn wagons delivered other foodstuffs from as far afield as Hereford and Carmarthen. Four-horse millers' vans brought in sacks of flour from the Velindre Mills in Brecon, all stored in large warehouses at the rear of the main building. The shop had its own bakery and butcher's department. Drovers rode in with herds of animals which were kept in pens before being taken to the shop's own slaughterhouse behind the stockade. Altogether some fifty different tradesmen were employed, and branch premises were set up in surrounding areas.

By the early 1850s when Buchan had formed his own company, Andrew Buchan & Co., to run his wide business interests, the man was already becoming a local legend. It was said that when he saw a small boy with a shabby cap, he would toss the ragged headgear over a hedge, take him into the shop and give him a new cap along with a bag of biscuits. He presented each member of the Rhymney cricket team with a gold sovereign each time the team won. The stories grew in size. During a time of famine, it was said he loaded a ship with iron from the works, which he sold in America, sailing back with a cargo of food.

Perhaps more characteristic of his firm grasp of business was another tale. A number of drays, loaded with beer barrels, had been left near the shop. One night some workers broke into the casks and drained the contents. A meeting was held with all the men to discuss the incident and rather than call in the police it was agreed that a weekly stoppage of a halfpenny should be taken from their wages. This was done – and the men repaid the debt over and over again.

Andrew Buchan died in 1870, aged eighty-two, leaving a fortune of over £50,000. He was buried in Rhymney and a tablet in his honour in St David's Church fittingly described him as 'Merchant of this Parish'. Though he gave his name to a brewery, he was always much more than a brewer, a larger-than-life figure who dominated Rhymney. Commemorative mugs, appropriately sold in his shops, were made to mark his passing. A contemporary remarked: 'Andrew Buchan was a name to conjure by in Rhymney and a wide district around for 40 years. He lived to a good old age and was respected by everyone.'

A succession of managers continued at the shop and brewery,

GLAZED IMAGE: a mug sold in Andrew Buchan's famous shop celebrating the man – and his passion for snuff (Welsh Folk Museum)

the most significant being William Pritchard who took over in 1874. He had to struggle with a sharp decline in the iron industry in the 1880s. In 1890 the Rhymney ironworks were closed down. He also had to contend with mounting criticism of the truck system. Legal proceedings were taken against the shop and the system was brought to an end in 1885, the company promising 'there shall be no more truck at Rhymney'. The store was probably the last 'tommy shop' in Britain, though it continued trading until 1911.

Burdened by these problems, William Pritchard 'in a fit of depression' took a gun, went behind the railway wagons in the brewery siding, and shot himself in 1898. The *Brewers Journal*, not wanting to recount the details, simply said that he died 'under very distressing circumstances'. He was sixty-five.

In 1901 Andrew Buchan & Co. was still lengthily advertised as 'brewers, wine and spirits merchants, farmers, grocers, provision dealers, butchers, drapers, jewellers, ironmongers, furniture dealers, cycle agents and hay and straw dealers' – but it was the brewery that was by now the dominant business.

'Considered the largest in South Wales' by 1867, it was brewing 12,500 barrels a year in 1878 and owned twenty-nine tied houses. In 1888 Buchan's Old Beer won a silver medal at the Brewers Exhibition in London, with their stout taking a bronze. The Rhymney Brewery also boasted a gold medal from a continental competition, and in 1902 first brewed King's Ale – 'The Wine of the Valleys' – for Edward VII's coronation. Besides the brewery's 'prize-medal ales and stout' its houses were also known for their nips of BOBS and BOBI – Buchan's Old Blended Scotch and Old Blended Irish.

TIME-HONOURED TRADITION: silver watch presented to drayman Thomas Williams for the best three-horse team at the annual parade in 1901 (Mike Williams)

William Pritchard's nephew Francis took over the management following the death of his uncle, but then left in 1900 having bought the Western Valleys Brewery at Crumlin – a company which was later to be brought into the Rhymney fold.

The brewery was by this time an impressive concern, dominating the bottom of the valley alongside the railway line. Particularly striking were the unicorns – teams of three horses used to pull the drays, with two side by side in the shafts and one in front as leader. Altogether over forty Clydesdales and Shires were stabled at the brewery.

Pulling heavy barrel-loads up and down the steep valleys, they needed experienced hands and strong arms to handle them. The draymen were paid 18s. 6d. a week with a low-rent, brewery-owned house and a free gallon of beer a day poured in. During the summer the town enjoyed the annual parade when the well-groomed horses, some wearing straw bonnets and tinkling bells, would march past, brasses gleaming. A keenly-contested prize of a silver watch was presented to the best turned-out team.

Francis Pritchard was succeeded by Captain Thomas Edwards who triggered a policy of expansion. In 1902 one of Pontypool's main breweries, the Crown Brewery, was taken over with fourteen pubs, extending trade away down the valley to the south-east as far as Usk. Buchan's was able to pick up the business for £9,000 as Crown was in liquidation, tragedy having also struck the Pontypool firm.

The company had been formed in 1896 with a capital of £10,000, the main partners being G. Paxton, the previous owner of the brewery, and Daniel Lewis of Mulberry House, Mamhilad, who owned several hotels and pubs. Mr Lewis was the managing director of the new concern. The partnership did not last long, as the *Brewers Journal* grimly recorded in

1899: Daniel Lewis 'was walking across the metals near Pontypool when he was knocked down by a shunting train and terribly mutilated. Medical assistance proved in vain, and the unfortunate gentleman died shortly afterwards' (on 22 April).

The brewery in George Street seems to have been dogged by misfortune. When owned by Sackville Lupton in 1879 the premises 'suffered severely by storm, above five feet of water accumulating in the brewhouse, stable and other portions of the premises, and the side wall of one of the buildings giving way, bringing the roof with it', reported the *Brewers Journal* in September. 'A valuable horse nearly drowned in the brewery stable and no less than 80 barrels of beer were washed out of the building and their contents wasted.' The damage, estimated at £400, helped Lupton go bankrupt in 1881.

With the wider trading area Buchan's found it difficult to meet all its deliveries by horse and so bought its first steam vehicle in 1908 with a top speed of 8 m.p.h. Soon two more were added to speed deliveries of its draught beers. Its basic brews were XX and stout with a stronger Golden Hop and IPA. In the same year long-service medals were introduced for loyal workers.

It was just as well the company invested in steam vehicles, as on the outbreak of war in 1914 many of the finest dray-horses were loaded into horse-boxes at Rhymney Station bound for the front lines in France, never to return. At this time Buchan's also bought the Blast Furnace Inn in nearby Pontlottyn, which had a small fourteen-barrel brewery attached.

In 1920 control of the parent Rhymney Iron Company – which since the closure of the ironworks in 1890 had concentrated on coal mining – passed to one of the leading mining companies, Powell Duffryn, who soon decided that it had little taste for the brewing business. In 1923 and 1924, as the minute books reveal, the company tried to sell to Wales's

LOADED UP: a Rhymney dray ready to roll in the 1920s. Note the Andrew Buchan clock at the brewery gates (Whitbread Archive)

leading brewers, Hancock's of Cardiff and Swansea. The negotiations failed. In 1925 they were in discussion with an English company, Wintle's Forest Steam Brewery of Mitcheldean in Gloucestershire, about a possible cross-border amalgamation. Again the talks came to nothing.

As trading conditions worsened, Powell Duffryn pushed harder and harder for a resolution. In 1927 and 1928 detailed discussions took place with Webbs Brewery of Aberbeeg only to collapse at the final fence. A disappointing meeting with Newport brewers

Riding the
Hobby-Horse

Buchan's gained a valuable asset when it took over Pritchard's in 1930 – a distinctive trademark that came to symbolize the brewery throughout its trading area.

A sporting member of the Pritchard family had tired of the cluttered old crest on the Western Valley Brewery's notepaper, which showed a wheatsheaf,

a prancing horse and swords and spears wrapped in a standard, all displayed on a shield surrounded by fruit, flowers and shrubs above the motto

'Purity and Excellence'. It was not the sort of symbol to grab instant attention.

So with a smile on his face he created the famous hobby-horse, showing a red-coated huntsman astride a barrel-bodied nag with jointed legs. The strange beast caught the public's imagination and Buchan's was swift to jump into the saddle when it bought Pritchard's. Their territory became known as the land 'Where the Rhymney Hobby Horse Roams'.

As a marketing move it was decades ahead of its time. It created a bond of affection between the old iron company brewery and its customers

Lloyd and Yorath early in 1929 forced the directors to conclude that there was no hope of a merger with one of the larger Welsh companies – at least not on the terms they wanted. So they considered 'coming to an arrangement with some of the smaller privately owned breweries in the neighbourhood'.

This policy led them to the door of Griffiths Brothers of Blaina. The Griffiths ran a brewery in High Street, Blaina, but more significantly an estate of twenty-eight pubs, some in Newport where they also operated a depot in Alexandra Road. A merger would give

Buchan's an important opening into Newport's trade.

On 9 October 1929 the two brewing businesses, valued at £450,876, were combined under the title Andrew Buchan's Breweries. The first chairman was Lieut-Colonel G.L. Hoare. Mr G.L. Pares, who had managed the brewery since 1915, also became a director. Rhymney had finally become an independent company. The old iron link had been broken at last.

The changes ushered in a new spurt of acquisitions. Within nine months Buchan's had bought the goodwill and most of the pubs of Pritchard's Western Valleys Brewery at Crumlin. This brewery, beneath the famous railway viaduct, had been bought by Buchan's former manager Francis Pritchard in 1900 and was registered as a company, D.F. Pritchard Ltd, five years later. Its estate extended to Merthyr Tydfil when Pritchard took over Christmas Evans' Heolgerrig Brewery in 1916.

Besides the houses, Buchan's also inherited a prize-winning brew from Pritchard's, Empire Ale. Its draught beers were now Empire, GHB (Golden Hop Bitter), BB, IPA and Stingo; with Light Ale, Empire, Special, Cream Stout and Family Stout in bottles.

The Depression slowed further advances until 1936 when their new emblem of the hobby-horse suddenly burst into a gallop, taking over two major breweries in a year. The

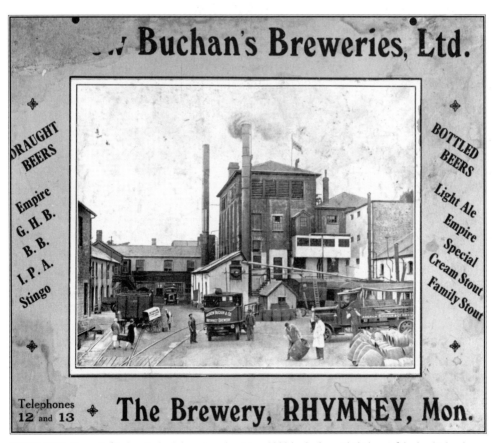

INSIDE VIEW: the yard of Andrew Buchan's brewery as shown on a 1920s' calendar, with the brews of the day displayed (Welsh Folk Museum)

first was David Williams' Taff Vale Brewery with some thirty pubs, the last sizeable brewery left in Merthyr Tydfil.

Buchan's claimed there had always existed 'a keen but friendly rivalry' between the two companies 'which spirit undoubtedly had a bearing on the eventual amalgamation'. As a reflection of this relationship, Taff Vale's chairman and managing director, Captain J.D. Griffiths, joined Buchan's board and was appointed general manager of the group.

The other acquisition was a more ambitious affair, taking Buchan's into the heart of the Welsh capital when it gained a controlling interest in Crosswells Cardiff Brewery. With this significant purchase the hobby-horse had cleared a major hurdle and moved into a bigger brewery league.

Crosswells history is almost in a league of its own. The name came from a brewery in the West Midlands – Crosswells Brewery of Oldbury, near Birmingham, run by Walter Showell & Sons. This ambitious firm, registered in 1887, began to market its beers across a wide area, opening depots in Bristol, Newport and then in the early 1890s in Cardiff. The stores in Penarth Road, trading under the title Crosswells Ltd, sold not only Showell's ales but also leading brands like Bass and Guinness besides wines and spirits. Much of the beer was bottled on the premises.

In 1897, tiring of shipping beer down from the Midlands, Showell's formed a new company called Crosswells Cardiff Brewery with the aim of taking over and combining the businesses of the Caerphilly and Castle Brewery, Caerphilly, with maltings and eighteen pubs; Crosswells Ltd of Cardiff and Bristol; wine merchants Carey & Co. of Cardiff, and four hotels – the Lord Wimborne, Ruperra and Avondale in Cardiff and the Wingfield, Llanbradach. The share capital was £140,000 and the directors, under the chairmanship of Walter Showell junior, included Alderman P.W. Carey of Cardiff, a leading figure in the licensed trade, and James Hurman, director of the Caerphilly and Castle Brewery.

It was a substantial combination – the properties were valued at over £250,000 – and the aim was to 'further develop' the Caerphilly and Castle Brewery, the only brewing company left in the town following the merger of the Caerphilly and Castle breweries in 1893. The premises off Nantgarw Road were described as 'extensive', but there were major problems with the site or buildings for almost immediately the new company abandoned their plans for Caerphilly and decided to build a completely new brewery elsewhere.

As managing director Frederick Richards later recalled:

In taking over the management of the company the directors found that the original intention to extend the brewery at Caerphilly would not be best serving their interests for several reasons, and it was therefore decided to look out for a more suitable site. After lengthy negotiations the directors arranged to take a plot of land near Ely Station.

Someone had made an expensive mistake. An EGM had to be called in March 1898, less than six months after the inaugural meeting, in order to increase the capital to £200,000 to meet the extra cost of the new brewery. There was no money for fancy flourishes, as planners Arthur Kinder & Son of London explained:

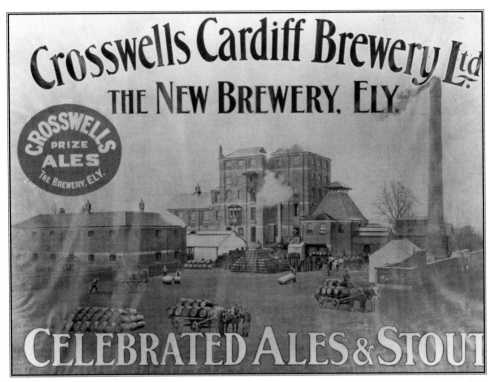

PROUD POSTER: Crosswells proclaim its 'celebrated ales' shortly after the New Brewery opened in Ely in 1900 (Whitbread Archive)

All the buildings have been kept as plain as possible to avoid any unnecessary expense on external design. Internally everything has been arranged with the greatest care to ensure the trade being conducted on the most economical and effective lines.

The brewery was not completed until late 1900 and the delay seriously hampered the enterprise. Instead of brewing their own beer they had to ship some four thousand barrels a year down from Birmingham at a carriage cost of 4s. 6d. a barrel. Trading profits in each of the first two years were just over £13,000, rising to £21,929 in 1900. No dividends were paid in 1898 or 1899.

Crosswells never recovered from this poor start. After optimistic noises at the beginning of the century, the brewery fell into a trough. Dividends on the ordinary shares were again not paid for years from 1904 and even the payments on the preference shares fell into arrears, only beginning to be paid off at the end of the First World War. The company's symbol was a beehive but there had been precious little honey.

Only in 1927 did Crosswells feel confident enough to make a major acquisition when it bought the central Cardiff brewery of William Nell with twelve pubs. Its only other significant purchase had been the small Tonypandy Brewery based at the Dunraven Arms in 1899.

Nell's Eagle Brewery had been established in St John's Square, beneath the tower of the famous church, since 1846, but it was always best known for its wine and spirits cellars

GRAND DESIGN: the plans for Crosswells Brewery were kept simple — but the premises were large

rather than its beer. The Nells were noted importers and dealers, particularly in Hungarian wines, and operated a healthy trade in supplying ships. At times the family firm behind the Tennis Court Hotel was only listed in trade directories as wine merchants. Nell's also ran a branch establishment at Bridgend, including a maltings.

On the death of William Nell in 1871 the business passed to his son William Walter Nell, who in 1890 registered it as a limited liability company valued at £65,000. The price was not a reflection on the brewery. Much of the value was hidden below ground in the extensive stocks of rare brandies, ports and champagne. Eight years later the company bought the small Abergwawr Brewery at Aberaman, near Aberdare.

In the early 1920s the company hit leaner times. In 1926 Nell's made a profit of just £1,361. No dividend was paid. In the same year two directors died, including Edmund Edwards who had been with the firm for twenty-one years. Nell's looked ripe for takeover and with Major John Griffith Jones on the board, it was expected that the Ely Brewery would take control; Griffith Jones was also chairman of Ely. Instead Crosswells swooped to thwart its Ely rivals.

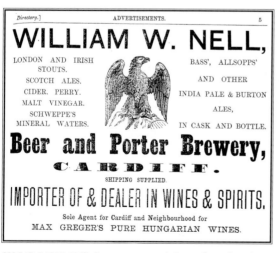

SPREAD EAGLE: Nell's Brewery was not only known for its beers but, as this 1875 advert shows, was a noted wine and spirit merchant

Nine years later the two neighbouring breweries, who stared at each other across the main railway line at Ely, on the outskirts of Cardiff, were being linked again. 'According to Stock Exchange reports, negotiations have been proceeding for an amalgamation of the Ely Brewery Company with Crosswells Cardiff Brewery', the *Western Mail* revealed in January 1935.

If the rumours were correct, the talks came to nothing. Instead Crosswells AGM in October 1936, at their St Mary Street offices, turned into a game of musical chairs. 'The chairman [Lord Glanely] in the course of his address to the shareholders announced that in consequence of the control of the company having passed into other hands neither he nor Mr C.H. March offered themselves for re-election and that Mr W.H.P. Rees is also resigning from the board as the purchasers will be electing a new board of directors', reported the press. The only one to stay on was Mr J.W. Walmsley who was joined by Buchan's chairman Lt.-Col. G.L. Hoare and Mr G.D. Kemp-Welch. The men from the valleys had moved in.

In the same year, Rhymney was able to celebrate its successes with an historic royal visit. King Edward VIII called into the old company shop – which was now being used for various work projects – to talk to the unemployed. Workers sang 'Cwm Rhondda' to him. The new king was moved by his tour and, unprecedented in those days, spoke out about the need to find jobs. It was his last public appearance before he abdicated.

Buchan's was to make one more takeover before the Second World War, when in 1939 it bought the Reform Brewery of Abersychan, just to the north of Pontypool. The brewhouse was old and small, the premises dating back to the Reform Act of 1832 after which they were named. But the tied estate was larger as the company had taken over the pubs of the ill-fated Westlake's Brewery of Cwmavon.

Westlake's was a castle-in-the-air concern. It had started humbly enough when Charles Francis Westlake bought a small brewery in Blaenavon in the early 1880s. In 1889 he formed a limited company, Westlake's Brewery Limited, with Michael Clune and Col. F. McDonnell, with a nominal share capital of £50,000. The company was valued at £35,000 and the profit that year was £3,245.

There was just one problem; a serious one. The water supply at the brewery was unreliable. In warm summer months the beer had to be brewed in Bristol, where Westlake's held its annual meetings. To overcome this dry difficulty the company decided to build a new brewery lower down the valley on a well-watered site at Cwmavon. But this was not to be any brewery; it was a dream in stone.

The grand 24-quarter edifice was built in 1900 by leading brewery architects, George Adlam & Sons of Bristol. 'The construction of the building is of the most substantial character in every way', enthused the *Brewers Journal*. 'The plant will be of the most modern description, both scientifically and practically.'

It nestled in a picturesque stretch of the valley like a great cathedral of beer – but would it ever be able to attract a large enough congregation to worship? Westlake's tried.

Now under the management of Charles William Westlake, it promoted its 'nourishing ales brewed from the best quality materials and pure rising spring waters' up and down the valley. By 1907 it had eighteen pubs and in 1905, 1908, 1909 and 1910 its beers won medals at brewing exhibitions. In 1911 the company took over the Castle Brewery in George Street, Pontypool, with nine pubs.

CATHEDRAL OF BEER: a sketch of Westlake's grand brewery at Cwmavon, as it was planned by brewery architects, Adlam & Sons of Bristol, in 1900

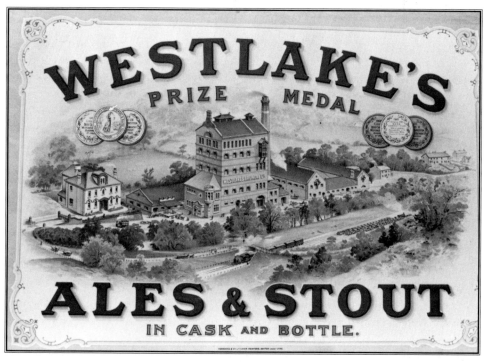

PRIZE ALES: a poster of 1910 proclaims Westlake's latest medal successes (Whitbread Archive)

Then the dream was shattered. In the depressed trading conditions of the late 1920s the company collapsed. In January 1929 the directors George Anstee, Edward Thomas and Col. Philip Pennymore signed over control to Thomas Skurray, chairman of the Brewers' Society and a director of many brewing companies including Hereford & Tredegar.

Brewing had ceased at the grand edifice in 1928. It was too big for the times — but that was not the end of its place in history. The towering buildings were taken over in 1935 by another group of people with a vision. The Eastern Valley Subsistence Production Society aimed to alleviate the mass unemployment of the Depression. Inspired by a Newport Quaker, Peter Scott, the society provided work in food production, clothing and agriculture. There were no wages but the workers could buy the goods and food they had helped make at much reduced rates. The scheme continued until 1939 when war solved the unemployment problem.

Today the Westlake's Brewery still stands in the valley, one of the few old brewery buildings in Wales not demolished, though it looks rather forlorn behind a sign proclaiming 'Extrusion and Moulding Compounds (Plastic Processors) Ltd'. How are the proud fallen.

The Westlake's pubs were eventually absorbed into a new company registered in July 1933, called the Reform Brewery, Abersychan. This was formed in conjunction with Daniel Seys Davies, once licensee of the Lion Hotel, Union Street, who ran the Abersychan Brewery from the original Reform Brewery premises alongside the pub.

The new company felt there was still some magic left in the old name since they carried on trading under the Westlake's banner. A letter to Elgam Workmen's Club in Blaenavon in 1933 boasts of the heady virtues of Westlake's draught XXB (at 4d. a pint), XXX and IPA, besides its three bottled stouts and a dinner ale.

In 1939, when Buchan's took over the Reform Brewery, it celebrated a hundred years of brewing at Rhymney. It owned an estate of 362 pubs and trade stretching from Gloucestershire to Pembrokeshire. The authorized share capital reached the magic figure of £1,000,000 in June. The company had moved far beyond just slaking the thirst of local ironworkers.

The war saw the Rhymney offices serve as the headquarters of the West Monmouth Home Guard under the command of the managing director. The company lost their vice-chairman when Capt. G.D. Kemp-Welch was killed in action in 1944. He was replaced on the board by his brother Mr P.W. Kemp-Welch.

Buchan's trade was soon extended following the conflict when it bought houses in Wales from the Cheltenham Original Brewery, after the Gloucestershire company had taken over the Hereford and Tredegar Brewery in 1945.

The Tredegar Brewery, founded by William Robins in Church Street before 1835, was unusual in having combined with an English brewery. In 1899, under John Jenkins, the company joined forces with George Edwards of Hereford to form the Tredegar and Hereford Brewery Company, as part of a trading agreement with the large Burton brewers, Allsopps. The power obviously soon swung across the border as three years later the firm's title was switched round to the Hereford and Tredegar Brewery. Brewing ceased in Wales. In 1905 the company absorbed the Brecon Brewery.

Buchan's gained more houses when it bought another brewery in 1945, Charles Edwards of Llanfoist, near Abergavenny, with some fifty pubs. The purchase added an extra arm to

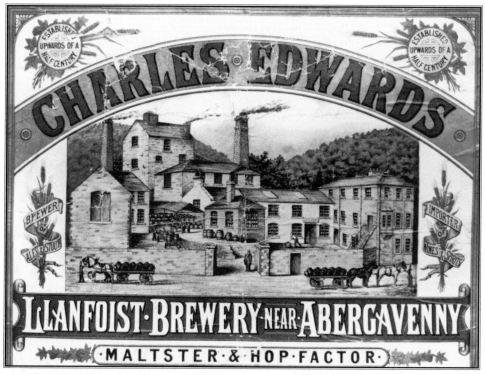

VILLAGE GIANT: 'The principal business of Llanfoist is the extensive brewing and malting establishment of Mr Charles Edwards', said Slater's 1868 Directory. Judging by this poster they were right (Whitbread Archive)

the business. The substantial brewery premises that dominated the village were converted in 1949 to the production of Llanfoist Table Waters, producing Buchan's own brand of soft drinks.

In 1951 there was a much more significant development, when the hobby-horse came on intimate nodding terms with the hind's head. Rhymney's close association with Whitbread dates from this year, when Colonel Bill Whitbread and Jack Martineau of the London brewers joined Buchan's board.

Why Rhymney made this crucial connection – which was later to lead to their takeover – is a reflection of the changing post-war atmosphere in the industry. Brewing had always been a friendly business; competition might be keen but everyone was generally happy to rub shoulders over a drink. They all stood together against the anti-alcohol lobby and government taxes and restrictions. At times it was too cosy; price rises were always well orchestrated.

Takeovers were usually agreed and often by invitation, a family firm wanting to sell up and enjoy the money rather than worry about the beer. Most of the leading families in the industry were well known, being jocularly known as 'the beerage'. After the war this close-knit community was shattered. Intruders were on the prowl.

The problem was property. Breweries owned a lot of it – in pubs, often on prominent street sites. And after the war property values began to rise rapidly. Breweries looked like potential goldmines. Developers like Charles Clore and Maxwell Joseph started to sniff around, looking for companies to take over.

Buchan's was feeling vulnerable. Its chairman for twenty-three years, Lt.-Col. G.L. Hoare, the dominant figure in the company since its founding in 1929, was leaving in 1952 to become a partner in his family's banking house, C. Hoare & Company.

Whitbread was well respected in the industry. It was the leading member of the beerage, the family company dating back to 1742. Col. Bill Whitbread, who had been chairman since 1944, was a formidably reassuring figure, who had successfully launched the famous firm onto the stock market in 1948. In 1953 he became the hero of the whole trade when, as chairman of the Brewers' Society, he stalled the Licensed (Amendment) (Tied Houses) Bill, which proposed to nationalize pubs on all new housing estates.

The London brewers had long sold their beer in South Wales, having had a depot in Cardiff since 1894. They also boasted some popular bottled brands, notably Whitbread Pale Ale and Mackeson Stout. It made sense to Buchan's to invite them in, let them take a share stake and seats on the board, and to sell their beer. Their arrival was reassuring, not a threat. The Rhymney Brewery was the first to seek shelter under what was later called the Whitbread umbrella.

There was also a second reason why the new chairman, Col. J.D. Griffiths, was delighted to have them on board. Whitbread was regarded as one of the leading technical brewers of

RATTLING SUCCESS: Crosswells new bottling line at its opening in 1958. Note close rival Ely Brewery peeping through the window (Western Mail)

the day – and Rhymney needed help. As a report prepared by the London brewers shows, the company was not running the most efficient plant.

Rhymney was dismissed as 'an old-fashioned brewery on the side of the hill which had a shortage of water in a dry summer. The fabric is old and leaking, but on the whole the equipment is not bad. The general impression created is rather Heath Robinson and definitely cramped quarters for much of the work.' Crosswells more modern brewery in Cardiff received higher marks.

The company must have been delighted when, in 1952, John Wilmot joined the board. He was the second brewer at Whitbread's Chiswell Street brewery in London. He was appointed brewer first at Rhymney and then Crosswells. Another Whitbread man, Frank Jupp, who joined the board in the same year, later became vice-chairman and managing director of Rhymney Breweries. Whatever happened in the end, at the beginning of the relationship the Welsh firm received good value.

Supported by Whitbread, 'Rhymney and Crosswells' – the joint name began to be used on pubs from 1952 – improved its operations. Brewing plant at Rhymney was renewed and the capacity of the brewery increased by the addition of new stainless steel fermenting vessels. The old iron company shop was converted into a bottled beer store in 1957 and a mineral water factory was erected on site, the premises at Llanfoist being closed.

At Crosswells Brewery in Cardiff a new bottling plant was opened in June 1958 by the High Sheriff of Glamorgan amid a fanfare of publicity. The modern filling machine could handle 400 bottles a minute, the three-storey extension more than doubling the amount of bottled beer that could be produced.

Significantly among those present on the grand day, which included lunch at the Angel Hotel, were Whitbread's technical director Mr A.H. Glenny and a new director of Andrew Buchan's – Col. Harry Llewellyn. Well known as an Olympic show-jumper, Col. Llewellyn became chairman a month later and within a year had changed the name of the company to Rhymney Breweries Ltd. A new 25-year trading agreement was signed with Whitbread in 1959.

A popular sportsman and dashing cavalry officer, Col. Llewellyn had been brought in to lead a charge into the massed ranks of South Wales breweries. Within a few months he had swooped to capture one of his main rivals, Ely Brewery.

THE PIT PINT

The Ely Brewery was a steady dray-ride outside the centre of Cardiff, just before the Ely Bridge, and the company spent much of its history building bridges between the town and the industrial valleys beyond. Appropriately its emblems were first the solid stone span at Ely and later the historic arch rainbowing over the Taff at Pontypridd.

The enterprise dates from the early 1850s when David Davies obtained a lease of land for a brewhouse at Ely from Lord Romilly. By 1863 Matthew Cross was brewing and malting on the site and, as sketches on trade cards show, the buildings were substantial for the time. Unlike their town-centre rivals, there was plenty of space for expansion in this semi-rural area, which was listed under Llandaff rather than Cardiff.

By 1875 James Ward was in charge of what he called the Tower Brewery, Ely. He boasted in an advert that he was brewer to both the Prince of Wales and the Marquis of Bute, carrying their coats of arms on his price list. He offered a wide range of three stouts, three

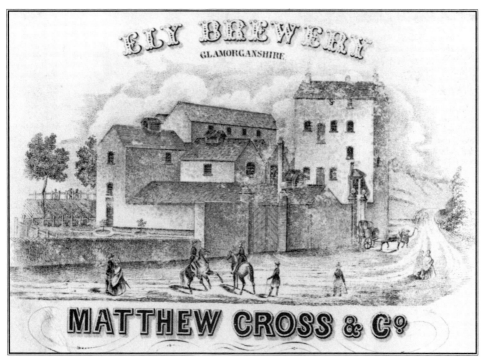

RURAL RETREAT: the Ely Brewery in the late 1860s when run by Matthew Cross, as depicted on his trade card (Welsh Folk Museum)

bitters and five 'not bitter' including his most expensive beer, XXXK OL at 30s. a kil (18 gallons), and one of his cheapest, XX at 18s.

In 1887 the firm was registered as the Ely Brewery Company Ltd and by the end of the century was brewing over 28,000 barrels a year. Chairman R.K. Prichard told the AGM in 1900 that the company had decided to pay the additional Boer War tax rather than reduce the gravity of its beer in order 'to keep up the reputation of the brewery'.

This was the year that Crosswells Brewery was built opposite in direct competition, proclaiming itself to be the New Brewery, Ely. The Ely Brewery retaliated with a large sign along its wall on the Cowbridge Road, stating it was 'the original Ely Brewery'. The pair made uneasy neighbours as each eyed the other up across the railway line.

Ely had by now bought a large number of pubs in the area including the eleven houses belonging to the Ship Brewery of Millicent Street, Cardiff, in 1899. Dividends on ordinary shares were running at 20 per cent a year. The successful company began to look up the valleys for more business, opening a depot in Commercial Street, Aberdare.

In 1920 this move up the valleys was taken a large stride further when Ely Brewery merged with Rhondda Valley Breweries of Treherbert and Pontypridd to form the Rhondda Valley and Ely Breweries Ltd. Eight years later this mouthful of a name was trimmed down to the Ely Brewery.

The Rhondda Valley Brewery of Treherbert had been registered in 1873 with a share capital of £10,000. Four of the first eight subscribers were innkeepers. In 1892 it bought the oldest brewery in Pontypridd, the Pontypridd Brewery in Taff Street, which had been established in 1853 by William Williams. Rhondda Valley Breweries was formed in 1896 to

BIG BUSINESS: a Rhondda Valley Breweries dray operating out of Treherbert at the time of the merger with Ely. Note the huge hogsheads (54 gallon casks) on the back of the lorry (Whitbread Archive)

combine the two businesses under the chairmanship of George Evans of Plasydderwen, Pontypridd.

This firm was no small beer as it already owned or controlled ninety-seven pubs in the heart of the thirsty South Wales coalfields. Its assets in 1896 were valued at £237,000 and the average annual profits for the previous three years were almost £20,000. In 1897 the new company bought the Rhondda Valley and Pontypridd Bottling Company and acquired the valuable district agency for Burton brewers, Bass.

In 1918 Rhondda Valley Breweries further strengthened its position in the valleys by taking over its local rival, Pontypridd United Breweries, which had been formed in 1903 through the merger of David Leyshon's Graig Brewery in Rickard Street and the Newbridge-Rhondda Company, Glen View Brewery, Court House Street.

This was the sizeable valleys' venture that merged with the Ely Brewery two years later, and although the Ely name soon came to dominate the firm, the vast majority of the familiar blue and white houses were not around Ely but up the valleys. Its strength lay deep in the coalfields rather than in Cardiff. The combined concern was said to be 'one of the largest and most progressive companies in Wales and the West of England'.

Though the three breweries (production in Pontypridd having been concentrated at the Glen View Brewery) were on short time during the coal strike of 1921, trade soon backed up this claim. Profits in 1922 edged in front of the Welsh brewing giants Hancock's (£55,418 to £52,982) and then rocketed ahead to £91,463 in 1923 and £131,984 in 1924. Hancock's only managed £50,111 and £65,876 in these same years. The company also benefited from owning almost all the share capital of the Cardiff Malting Company.

Miners were heavy drinkers and so a brewery directly meeting their needs (eight out of ten Ely pubs were up the valleys) could make staggering money. The trouble with this heavy reliance on one industry was that if the miners went on strike or were put on short-time working, profits fell as fast as a stone tossed down a pit shaft. In 1926 came the General Strike which paralysed the coalfields for eight months.

Chairman John Griffith Jones told the AGM in 1927 that 'owing to the abnormal depression of trade through which the company has just passed and the slow recovery', they were unable to declare a dividend. A Treherbert licensee at this time, under a headline 'Pub crisis', claimed that 'practically 90 per cent of the Rhondda publicans are on the verge of bankruptcy'.

This was crippling news for a brewery which had 284 pubs chiefly in the Rhondda and Aberdare valleys in the heart of the Glamorganshire coalfield. Not to mention three large breweries in Treherbert, Pontypridd and Ely. Drastic surgery was applied. The Treherbert brewery was closed in 1928, the one in Pontypridd a year later. All production was concentrated at Ely and the company's name changed to the Ely Brewery.

The company lost its energetic chairman Major John Griffith Jones when he died in October 1928. A remarkable man, he was not only an accountant, soldier, Freemason, golfer and huntsman, but also a member of the governing body of the Church in Wales. He was the son of an even more remarkable man, Caradog, the celebrated Welsh musician who with his 'Cor Mawr' laid the foundations of the Welsh miners' fame in choral singing.

Worse was to follow in December 1928, with shareholders' allegations of mismanagement. The brewery troubleshooter, Thomas Skurray, chairman of the Brewers'

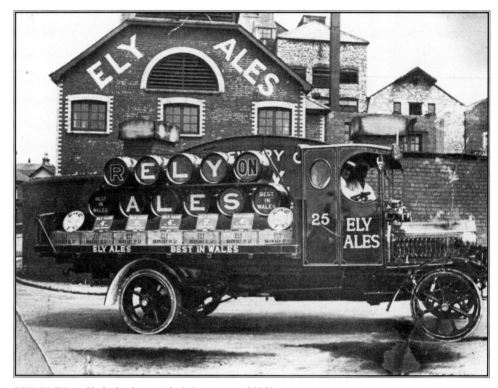

RELY ON ELY: an Ely display dray outside the brewery around 1930

Society, was brought in as managing director to steady the rocking ship. But even his arrival could not prevent a loss of £21,918 being declared at the 1929 AGM.

With the deep depression of the early 1930s hitting the coalfields hard, Ely had little chance to recover. Every chance to cut costs was taken, the bottling stores and offices in Taff Street, Pontypridd, being closed in 1930 and the staff moved to Ely. In 1935 there was talk of a merger with their close rivals Crosswells, but nothing happened. The *Investors' Chronicle* commented that year that anyone buying Ely shares 'is drawing a very long bow, but like the man who buys a depressed foreign railway stock, may be consoling himself with the thought that at any rate things cannot get much worse'. The company reported a slight improvement in trade in 1936 – the loss for the year of £8,650 was less than the loss for the preceding twelve months (£11,431). The company was not so much on the floor as nearly underground. Dividends had not been paid since 1927, and Ely was deeply indebted to Barclays Bank. The total debt in 1936 was £105,654.

Chairman Thomas Skurray died in 1938. His eminent position within the brewing industry – he was chairman of the Brewers' Society from 1929–31 – probably helped keep Ely afloat, but his heavy responsibilities elsewhere meant he was unable to devote much time to the detailed running of the brewery. His home was in Berkshire where he was a former chairman of the county council. He was also chairman of two other breweries, Morlands of Abingdon and Wilsons of Manchester, besides being a director of other companies. He was deeply involved in the restoration of the Trust Houses group of hotels.

The man who eventually replaced him at the end of the war was to have much more devotion to local detail. He almost became Ely. He also restored the fortunes of the brewery. His name was Lazarus Nidditch.

The war was not bad for everyone. Increased demand for iron and coal coupled with a return to full employment meant that an odd prosperity, which had evaded the years of peace, returned to South Wales. Brewing materials – malt, hops and sugar – might be in short supply, but the demand was there. Ely's advertising began to take on a fresh bounce. 'There isn't much but what there is of it is good', said one wartime slogan. 'We know it's scarce, but it's worth fighting for', added another.

Speculators even began buying Ely stock in the belief that there was at last some light at the bottom of the pit shaft. They were paying 3s. 6d. for £1 shares in 1944. The *Sunday Express* described Ely as potentially 'a £2,000,000 company'.

But the brewery was still a long way from recovery. Chairman A.F. Pollard starkly illustrated this in 1944 when he revealed that the firm was only operating at 38 per cent of its peak production in the 1920s. And much of the present trade was due to the war. He believed they needed to reorganize the capital structure and relaunch the company with a new board – and the man to carry this out was Lazarus Nidditch.

Mr Nidditch was appointed financial director in December 1946, and two months later became chairman. He took on the additional role of managing director in 1949. He hit the depressed brewery with all the force of a runaway dray lorry hurtling down a steep hill. Everyone told him he had no chance of turning the ailing concern around. He refused to listen – and surprised them with the unexpected. Like a dividend.

The first for twenty-one years was paid in 1947. It was only a modest 2½ per cent, but it was a major achievement for Ely. The next year the dividend was 10 per cent and profits were a booming £238,613. In response the value of the shares in 1948 increased to five times their 1947 level, and by 1951 the volume of trade was double that of 1939.

Lazarus Nidditch was an unorthodox financial wizard. But he was much more than just a miracle worker in the accounts office, as the *Joint Stock Companies' Journal* recognized in October 1953:

Without doubt much of the credit for a complete metamorphosis of both balance sheet and profit and loss account has to be given to this restless executive who gets through a tremendous amount of work, finding time to visit the inns and hostelries owned by the brewery throughout the Rhondda and South Wales, exercise his influence on customer relations generally, initiate and constantly develop a spirit of harmony and cheerful mutual effort amongst the staff and, practically gratuitously, act on the one hand as a highly competent valuer and on the other as a skilled taxation lawyer.

He made every effort to raise morale and not just among the shareholders. At the end of 1948 a bonus of three weeks wages was paid to all brewery staff. An Ely '25' club was formed the following year for long-service employees, and in 1952 a house magazine called *Mild and Bitter* was introduced. A welfare hall and canteen were built at the brewery and sports clubs were encouraged. Social outings were arranged for staff to places like

Henley-on-Thames. Workers were even given gifts 'such as shoes, overalls, woollen jumpers and blouses (for the ladies)'.

Throughout the 1950s steady programmes of improvements to the brewery were carried out. They were badly needed as the plant was worn out, most equipment being twenty to forty years old. Seven large fermenting vessels were fitted in five years, while the mash tun, copper and cooler were replaced or overhauled. A new £20,000 boiler room was installed in 1951, a laboratory built and the bottling plant modernized.

A scientist Dr Joshua Saper was brought in as technical director to assist the head brewer. Digby Leyshon, improve the beers. By Leyshon's death in 1953, Nidditch was boasting about 'the new reputation for our quality products'. He claimed at the AGM that year, 'Ely Brewery is in the forefront of breweries in the use of modern methods with modern plant of copper and stainless steel, instead of leaky and wooden fermenting vats 30–40 years of age, a worn boiler 55-years-old and other decrepit plant breaking down weekly.' By 1955 Ely was testing completely new technology, 'carrying out experiments in our laboratory for hop extraction by ultrasonic process' and 'studying the possibility of ozone treatment in the cleansing of casks and stoppers'.

Nidditch also left his mark on the 250 or more pubs, which he said were 'in a deplorable condition of disrepair through terrible neglect' when he became chairman. He should know – he visited 150 houses in his first year. He was determined to eradicate 'the distressing atmosphere existing in all houses until 1947'. He wanted to do away with the crude miners' bars with their 'vertical drinking' by introducing a more homely atmosphere with new saloon bars with comfortable seating. Helped by his wife Matilda, who was also on the board from 1951, he added such novelties as ladies' toilets 'to enable wives to accompany their husbands'. He was later proud to claim, 'We have given a moral uplift in almost every town and village between Cardiff and Rhondda.'

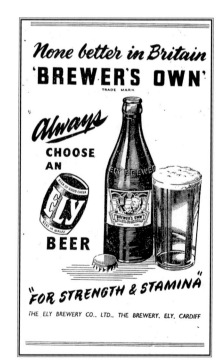

He was a keen innovator in every field, adopting the latest marketing strategies. A film was made in January 1948 depicting the brewery in operation, which was shown at local cinemas. The old logo of Pontypridd bridge, which he dismissed as 'far from exhilarating' and bearing 'no relation to our products', was replaced by a large Ely barrel. Red, white and blue electric lanterns were fixed to every pub 'for easier night recognition of an Ely house'.

He was conscious of the need for strong presentation. A stream of new beers appeared, three in 1948 alone spearheaded by Brewer's Own, a bottled pale ale, which was sold with lines like 'Fit for a Golden Tankard' and 'Best in Wales

AIMING HIGH: a 1950s advert for Brewer's Own, the pale ale Nidditch hoped to turn into a national brand in Britain

– None better in Britain' accompanied by a capped brewer gazing in admiration at his glass on the colourful label.

His sales horizon did not stop at the Welsh border. Beers like Brewer's Own and Golden Gleam were advertised in the *Financial Times* and British trade directories. In 1947 bottled beer sales had been minimal; ten years later a range of bold brands, rivalled only in presentation by Flowers Breweries in Britain, accounted for 15 per cent of a much increased production.

Many brewers were worried about the growing challenge of television, which was threatening to stop customers going to the pub at night. Nidditch typically tackled the problem head-on by launching a TV Ale in two-pint flagons in 1953 with the slogan 'When you're watching Wenvoe [the BBC broadcasting mast near Cardiff] you can still get Ely service.' He even composed a ditty to accompany the new beer:

> We planned long hours of viewing
> When TV came to Wales;
> Our glass we'd keep renewing
> And watch if Harding fails.
>
> These dreams now need reviewing
> Because of rising sales,
> We're much too busy brewing
> Large stocks of TV Ales!

Nidditch liked catchy lines. On the back of the 1948 annual report was printed: 'There are no hopeless situations – there are only men who have grown hopeless about them' and 'It's easy to avoid criticism – say nothing, do nothing, be nothing.' The messages had a hard edge and Nidditch did not suffer men he believed were failing to pull their weight.

He was the commander and the men were his troops to be moved around at will. His manipulative attitude to his workers is best summed up in his candid report to the 1951 AGM:

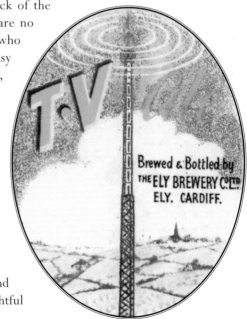

I submit that I understand the outlook of working men and human material at all levels of intelligence, and can assess and place square or round pegs in their rightful and most advantageous position, where their temperaments and flairs enable them to give of their best.

MIXED RECEPTION: the enterprising ale Nidditch introduced in 1953 to try and capture those drinkers who stayed at home to watch the new TV

His individual and autocratic approach, recalled one colleague, meant that someone showing initiative and coming up with a good idea could be promoted on the spot. On the other hand, someone crossing him in the morning could find the chairman himself unscrewing his nameplate off his office door in the afternoon. He had been summarily dismissed.

Lazarus Nidditch's shock tactics worked wonders in the company. He was also chairman of the subsidiary Cardiff Malting Company, in which Ely held a 36 per cent interest. In 1950 Cardiff Malting returned record figures. Outside his companies, however, his aggressive attitude raised hackles rather than profits. As early as 1949 he was in trouble for criticizing the Pontypridd licensing justices. But his main clash was with the vested interests of the mighty Stock Exchange, which he accused of talking down the true value of Ely's shares. His attacks were far from diplomatic, often being bitter and personal. He particularly despised the jobbers who handled shares, claiming they were 'a few men surrounded by privileges' who dealt in 'false prices'.

His arch enemy was one of the principal brewery share jobbers, Esmond Durlacher. 'King Esmond', as he was known, was a highly influential figure within the City. When the Red Tower Lager Brewery ran into trouble in 1953 he was brought in by worried shareholders to probe the losses at the Manchester company. He was not a man you could insult without sharp retaliation.

After some heated exchanges – Nidditch enjoyed circulating his shareholders with abusive attacks on the stock-market – Durlacher refused to handle Ely shares. The rest of the dealers followed suit, freezing Ely's shares out of the London market in 1950 and affecting the company's ability to raise vital capital.

This long-running sore caused some shareholders to challenge Nidditch's rule. The adoption of the company report was opposed at the 1951 AGM, but the chairman easily came out on top in a poll. The election of some directors was also contested in vain. With a remarkable profits record, he was impossible to topple.

By 1953 Nidditch had raised his gun sights and was attacking the chairman of the Stock Exchange, John Braithwaite, for failing to take action to get Ely share dealings going again. The City editor of the *Daily Express*, Frederick Ellis, made the mistake of expressing some sympathy with Ely's plight on 6 October, though he described both sides in the dispute as 'childish' and dismissed Nidditch's arguments as 'ravings'. The jobbers wanted Nidditch to apologize for his remarks before dealing could be resumed. Ellis thought the enemies 'should meet over a neutral pint' and sort out their differences for the sake of Ely's 3,000 shareholders. Immediately Ellis's City desk was hit by a thumping 578-word telegram. 'Mr Nidditch's telegram is an angry, violent, threatening and accusing document. Such is the fiery spirit with which he fights his case', Ellis reported the next day, 7 October. 'I spare you the details', he added. Ellis had no time for Ely's chairman or the 'elegant' Ronald Charles Quirk and Esmond Durlacher, whose jobbing firms dominated the brewery share market. They should resume dealing however abused and goaded, he advised.

The telegrams from Ely did not end there. On 8 October Ellis told his column's readers that he had received another '402 words of little wisdom'. On 9 October he reported that a further 562 words had arrived – which he had put in the waste-paper basket. Nidditch refused to accept the clear message that his expensive messages were not wanted. By the end of the year Ellis had awarded him a mock honour for services to the GPO for 'helping reduce the losses on the telegram service'.

The eccentric episode illustrates Nidditch's volcanic and single-minded nature. Another event in October shows a different side to his personality. Staff presented him with an onyx desk set prior to the 1953 AGM at the New Inn Hotel, Pontypridd, saying he had endeared himself to them through his 'never-failing kindness, consideration, good fellowship and sympathy'. He was a magnetic man who attracted devotion and dislike in equal measure. He was certainly impossible to ignore.

By the mid-1950s his long, rambling reports to the brewery's annual meetings were becoming more bizarre. In 1955 he launched into his view of the future, predicting cars would be driven by gas turbines in ten years' time and atomic power would be used to supply piped hot water to every home. His visions were interesting; their relevance to the company less obvious. One critic described them as 'a sometimes unintelligible tangle of personal views on irrelevant matters'. Some shareholders became restless, especially following one odd side venture – the purchase of the World Natural Sponge Suppliers (of which Nidditch was chairman).

There were also growing rumblings about 'Nidditch nepotism' since he and his wife had been joined on the board by their son-in-law (Dr Saper) and brother-in-law. The discontented found a voice when Cardiff accountant Julian Hodge began to use one of Nidditch's favourite weapons – a circular to shareholders – against him. One referendum voted against the decision to acquire the controversial sponge company. Another circular brought matters to a head in 1956 when it called for an EGM to establish a new board – without Lazarus Nidditch.

Mr Hodge did not mince his words. On 28 June he issued a parting shot headed 'Goodbye Mr Nidditch', listing eight reasons why the chairman should resign. The farewell was a trifle premature. Mr Nidditch came out fighting at the EGM on 9 July, attended by over two hundred shareholders. 'At times there was pandemonium, a number of men being on their feet and shouting in competition', reported the *Western Mail*.

Mr Nidditch survived the shouting match, a card vote for his removal being defeated by 2,564,856 to 2,028,157. He declared that his opponents had been 'well and truly licked' and accused them of mounting a 'muck campaign'. His support was particularly strong among licensees who had benefited from his revival of the brewery.

By the end of the year he had sold the main assets of the sponge company at a profit and stated in a colourful statement to the press that he was 'completely vindicated against the vile and bestial attacks made upon him between December, 1955, and July, 1956, that brought so much publicity and attempts to unseat him and bring his name into contempt'.

Mr Hodge, however, would not go away, continuing to make detailed criticisms of the running of the company in the name of Investors Protection Facilities Ltd, of which he was managing director. 'The battle of circulars . . . considerably increased and enlivened the mail of stockholders', reported the *Western Mail*.

But Mr Nidditch was in a strong position. In 1957 he said Ely's accounts showed the greatest percentage increase of profits, in relation to the number of pubs and capital employed, of any British brewery. He was also planning to open two motels and in 1958 bid

ONE OF THE LADS: Lazarus Nidditch (left) enjoys a drink at the brewery to celebrate a new cooper completing his training in 1958

for Gordon Hotels. He dismissed Mr Hodge's circulars as 'a serious abuse of the common rights of a shareholder for undisclosed aims'.

The criticisms continued – and hurt. The directors' report in 1958 claimed that:

> During 1957 there has been a continuation of the vindictive hounding campaign to stir up ill-will and prejudice against the chairman and bring his name and that of members of the board into disrepute and contempt for unlawful motives by attacks relying on fictitious and baseless allegations, the gross distortion of facts and figures and the assertion of half-truths.

Despite the war of words, business boomed. During 1958 a new bottled stout called Gold and Silver was introduced, followed by a keg beer, Silver Drum. The brewery was running so close to capacity, that until new conditioning tanks came into use these new brands were not promoted. The profits for the year were £269,981. 'I have obliterated despondency at Ely', declared Mr Nidditch in his favourite egotistical style, describing himself to shareholders at the AGM in February 1959 as the 'aggressive defender of your true investment'.

With half an eye on future developments, he added: 'I have said before and I say again that if anyone wishes to take over your company, considering the achievement resulting from virile leadership and management effort and success, they should do so in an honourable way by making an open bid at a fair price or leave us – and our shares – alone.' Lazarus Nidditch was tiring of the brewery business and open to good offers. In the summer he believed he had obtained one.

On 6 August 1959 a London finance corporation, H. Jasper & Co., made a £1.75m bid for Ely Brewery with its 260 pubs. The investment bankers had the previous month bought a Welsh chain of hotels, R.E. Jones Ltd, which included the Angel, Sandringham and Philharmonic in Cardiff, the Seabank and Esplanade at Porthcawl, the Mackworth in Swansea and the Piccadilly in London. Lazarus Nidditch immediately recommended acceptance.

His attitude was hardly surprising. If the deal went through he would receive £30,000 compensation for loss of office. His son-in-law Dr Saper, who was by now vice-chairman, would be given £10,000. Mr Nidditch would also receive £17,000 for his house, Two Pines, at Creigiau. The *Financial Times* was not happy with this arrangement. 'This is hardly good enough. If stockholders are being asked to buy out directors' agreements with the company – as in effect they are – they are surely entitled to know what they are paying for.'

More crucially, did Mr Nidditch really know who the buyers were? Harry Jasper had left Berlin for London in the 1930s, during the Nazi persecution of the Jews. His finance and property group only took off after 1956 when in three years he bought a whole string of companies including Cotton Plantations, Capital News Theatres, National Model Dwellings, Victory Real Estate, Temperance Billiard Halls, Blindells shoe shops, Rubens-Rembrandt Hotels, Reliable Properties and Lintang Investments. In each case the aim was rapidly to make money by selling or developing the properties involved.

'The identity of his backers has not been disclosed, though they are known to be a syndicate rather than an individual', commented the *Financial Times*, which estimated that in the few months before the Ely bid 'the group seems to have had available sums of the order of at least £15m to finance acquisitions'. Then the gravy train ran off the rails.

On 14 September everything seemed to be going smoothly. Mr Nidditch and the other Ely board members resigned at a special meeting and were replaced by four new directors including Harry Jasper as chairman and Friedrich Grunwald as managing director. 'I think I have done the right thing by seeing today that everybody connected with the company gets a fair and satisfactory deal', said Mr Nidditch.

Then the truth dawned. There were not enough funds to pay for the bid. As rumours spread, the value of shares in Jasper group companies slumped alarmingly, £6.4m being wiped off in three days. An embarrassed Mr Nidditch declared:

I have given an ultimatum to Mr Harry Jasper. He and his advisers all day failed to talk to me or see me or Ely Brewery's London solicitors as to payment to all shareholders which was promised by Monday or Tuesday. I have now challenged his policy and directorial appointments.

The next day, 18 September, just four days after resigning, Mr Nidditch resumed his position as head of the brewery, declaring that the election of Mr Jasper was void.

Mr Jasper claimed that the bid for Ely had been made on behalf of Friedrich Grunwald, the solicitor elected as managing director. However, he had now disappeared without providing the money. Mr Grunwald later turned up 'on holiday' in Israel, claiming to have had a nervous breakdown. On 21 September the Stock Exchange Council ordered that dealings in all fifteen Jasper group companies be suspended. On the same day Mr Nidditch accepted, following legal advice, that Mr Jasper was still in control of Ely Brewery.

The affair was a farce. As the *Financial Times* tried to explain: 'Mr Jasper's account of the matter is that Mr Grunwald, acting as a solicitor, made the offer on behalf of Mr Jasper, acting as a banker, who was in turn acting on behalf of Mr Grunwald in his capacity as a private client.' Would the real bidder please stand up!

There was also a much more serious side, as the *FT* revealed in its own editorial:

This is a blow to the City's reputation . . . in the City itself people will be well aware that this is a matter in which relative newcomers have played the principal parts; the public, however, cannot be expected to make close distinctions.

On 24 September Mr Jasper and his colleagues resigned from the Ely board and the takeover was officially called off. The Board of Trade appointed an inspector to investigate Jasper's affairs. Mr Nidditch added dramatically two days later, 'I believe a certain man will make a lightning bid for control. Things are moving at the moment. They are too fluid and dangerous for me to make any comment now which might embarrass or upset the situation.'

In the meantime he was prepared to step into the breach. 'I'm the best man to put things right', he told the *Empire News*. 'There are plenty of majors in the company now but no generals. I was the general. It took me 13 years work – sometimes 70 or 80 hours a week – to make it what it is now. If the shareholders choose anybody else then they must be prepared to take the consequences.'

On the eve of the shareholders' meeting, the headline-hitting events took another twist when two detectives working with the City of London Fraud Squad began inquiries in South Wales, interviewing Mr Nidditch on 29 September. The lengthy questioning continued until three o'clock in the morning. On the same day Leslie Lowe, the chairman of a Cardiff engineering company, was appointed temporary chairman of the brewery.

The packed shareholders' meeting on 30 September, which Mr Nidditch had convened, was a vicious vendetta, with many of the questions being about the £30,000 he had received as compensation for loss of office. 'If you sold your company for 30 pieces of silver, what are you doing here?' demanded one shareholder.

The *Western Mail* diary column, Wales Day By Day, best described the gladatorial contest in the Park Hotel, Cardiff:

From the moment of Mr Lazarus Nidditch's ceremonial entrance, to the craning of necks and the flash of camera bulbs, it was one of those vintage scenes, with raised hackles and stormy words, that one has come to associate with meetings of the Ely Brewery shareholders.

There was Mr Nidditch, a short, proud man with bald, nut-brown head, sitting behind the green-baize table, his rimless spectacles glinting under the chandeliers. On his left, the impassive, expensively-dressed figure of Mrs Nidditch – whose feminine touch once helped reglamorize the Ely Company pubs.

IN THE LION'S DEN: Lazarus Nidditch addressing the stormy shareholders meeting of 30 September 1959, supported by his wife Matilda (Western Mail)

Mr Nidditch would not claim that public speaking is his strong point. His words came in bursts of excitement and rhetoric. In spite of his inscrutable front, he is not a man who can conceal his emotions.

It was soon evident that the main struggle of the meeting was not so much between Mr Nidditch and the shareholders, but between Mr Nidditch and Mr Julian Hodge, the financier.

Yesterday phrases like 'fraud squad', 'double-cross', 'fait accompli' and 'completely untrue' fell like angry grape-shot from Mr Nidditch's lips. While before him, as though in the centre of the arena, sat a tight, confident group, headed by Mr Hodge, all well-armed with detailed documents.

At every awkward question from the floor, Mr Hodge – a handsome, upright, greying man, with all the poise of a seasoned warrior – would bare his gleaming teeth in a triumphant smile. And whenever he or one of his companions posed a question, Mr Nidditch would crouch forward and snap out a reply in a tone of undisguised vehemence.

It was not a happy meeting for Mr Nidditch. He was trapped and taunted. When a colleague proposed that he should be made chairman, he gave a resigned smile and replied, 'I am sorry, it is too late.' To complete his day, when he emerged from the bear-pit he found his car tyre had been let down.

Mr Nidditch had hoped to hold the reins of power again at the brewery. Instead a more accomplished horseman galloped in. Col. Harry Llewellyn, the famous Olympic show-jumper and chairman of Rhymney Breweries, saw his chance. On 21 October the hobby-horse brewers bought 900,000 Ely shares for £200,000 from Harry Jasper (which Jasper had ironically bought from Nidditch). This represented a stake of around 12.5 per cent. Rhymney was determined that its valley rivals should not fall into 'unfriendly and competitive hands'. On 29 October it followed up with a full bid, valued at £1.75m, offering Rhymney stock for Ely shares.

'The offer is in effect a merger of the existing stockholders in both breweries. If we are successful Ely beer – which is good beer – will continue to be brewed at Ely and sold in the Ely houses,' said Col. Llewellyn. 'This is not a takeover bid, this is a proposal to merge the interests of two established South Wales breweries, which will work out to the eventual benefit of both.'

The deal looked attractive to the leaderless company and its anxious investors. When temporary chairman Leslie Lowe outlined the offer at a shareholders' meeting on 4 November, he was applauded. When Lazarus Nidditch attempted to intervene with a point of order, he was greeted with shouts from all corners of the hall to 'sit down and shut up'. He had become yesterday's man – though he refused to leave the scene without a few fiery parting shots.

On 13 November he paid for adverts in the local papers welcoming 'the marriage of the two breweries' and offering advice on how to handle the merger. He concluded: 'I feel content that my stewardship and leadership of the last 12½ years has proved itself clearly to one and all.'

Four days later, in a circular to shareholders, his bitterness was apparent when he attacked Mr Lowe's handling of the shareholders' meeting – 'his much-vaunted schoolmasterly ability to keep his class in order' – and criticized the chairman's statement that the era of circulars was over.

Presumably he felt that his circular should be the last, but its terms are so unnecessarily provocative by its smooth imputations and its facts so distorted that it is itself the first of a new era of circulars because neither I nor the other directors who retired can be expected to allow its offensive, distorted and untrue comments to remain unanswered.

His circular ran to eight foolscap pages, vindicating his stance on every issue while sharply criticizing others. 'The Jasper failure was no fault of mine and was a shock to the whole country including many astute financiers.' He also questioned the legality of Mr Lowe's appointment and threatened legal action. He went out as he came in – with a bang.

The merger with Rhymney was completed on 16 December 1959, bringing together two large Welsh breweries to form the giant of the valleys. Ely had 260 pubs; Rhymney 470 – a grand total of 730 houses.

The combined company claimed to be the largest brewers in South Wales, with trade stretching from Fishguard to Hereford, where the appropriately-named Foxhunter pub was built. A flagship hotel, the Angel in Cardiff, was bought in 1962 and others in Swansea and Porthcawl run in conjunction with Fortes.

It soon became clear that it was not economic to operate two breweries – Crosswell's and the Ely Brewery – side by side in Cardiff. The trouble was neither was large enough to

DOOMED: Ely Brewery after the takeover by Rhymney Breweries. It only survived three years before being demolished in 1963 (Whitbread Archive)

absorb the trade of the other, so early in 1962 it was decided to build a modern extension to the Crosswell plant, virtually a new brewery, with offices opposite.

This development was officially opened in February 1963 by Lord Robens, chairman of the National Coal Board, who had a special interest in the new coal-fired boilers that burned 40 tons a week. The remodelled brewhouse was able to produce 200,000 pints of draught beer a day – 'the best round here' according to Rhymney's slogan – and handle 21,600 half-pints of bottled beer an hour, mainly Welsh Bitter, Brown, Strong and King's Ale plus Mackeson for Whitbread.

The Ely Brewery was closed and quickly demolished. Its barrel trademark and most of its beers were soon trampled underfoot by the rampant Rhymney hobby-horse. Lazarus Nidditch's pride and joy had disappeared and so had the man himself, leaving South Wales for Hertfordshire.

Capital Ale – It's Brains You Want

Visitors to the Welsh capital today might be forgiven for thinking that Welsh beer is all about Brains. On every corner in Cardiff there seems to be a Brain's pub, while the heavily-laden drays still rumble out of an arch from the Old Brewery straight into one of the city's main streets, St Mary Street. Railway bridges span the roads carrying the famous slogan, 'It's Brains You Want', and verses in honour of the city always manage to rhyme Cardiff Arms Park with Brain's Dark. But the name is a relative newcomer for a brewery that dates its history back to 1713.

At that time Cardiff was little more than a small walled settlement next to the castle, and the brewery would have been little more than a home-brew pub. The date is taken from a stone above one of the doors in the original brewery buildings.

The earliest Cardiff trade directory, Birds of 1796, lists no brewers at all, although there were eighteen inns. Brewing then was a small-scale domestic operation like cooking, not thought worthy of mention. The significant business was in the production and sale of the raw materials. The small town had six maltsters and hop factors, though four were part-time. Two were grocers, while another combined a more unusual mixture of trades. Elizabeth Jones was listed as a draper, hat dealer and maltster.

The first brewer mentioned was James Walters in Ridd's Directory of 1813. Predictably, he was also a maltster. The number of inns by then had jumped to thirty. His address is given as St Mary Street in 1822, and he was probably brewing on the site of what became known as the Old Brewery. This area, away from the castle, was ideal for brewing. A deep well had been sunk to reach the water-bearing gravel subsoils.

The problem with this position was that the River Taff at that time ran past the brewery in a grand curve on the opposite side of St Mary Street. The powerful river had already swept away St Mary's Church – skeletons from the old graveyard sometimes turned up on the mudbanks – and must at times have threatened to destroy the brewery. Flooding was commonplace.

By 1829 Edward Thomas Bridgen Carter had taken over the business – and the Old Brewery had a rival. John Wood's map of Cardiff in 1830 shows another brewery further up St Mary Street towards the castle, on the opposite side close to the Old Quay.

Pigot's Directory of 1835 was the first to list two commercial brewers in Cardiff, both in St Mary Street. E.T.B. Carter had gone from the Old Brewery to be replaced by Watson and Phillips and then, in rapid succession, Phillips and Andrews (1840) followed by William and Charles Andrews (1844). The latter were the first to use the name Old Brewery.

The other brewery was run by William Williams, who finally seems to have abandoned the Old Quay site in the mid-1840s and taken over the Old Brewery himself by 1848. On a map of 1849 the premises are referred to as Williams' malthouse.

A third commercial brewery had by then appeared to the east of the town in Great Frederick Street, whose founding family was later to play a part in the fermenting fortunes of the Old Brewery. John Thomas had established this rival concern by 1840, though within two years this New Brewery was being run by Elizabeth Thomas.

Then came the brewery boom. Cardiff in 1841 was still a small town, less than a third the size of Merthyr Tydfil. However, the opening of the Bute West Dock in 1839 and the completion of the Taff Vale Railway to Merthyr in 1841 opened up the riches of the valleys to the north. Iron and coal began to flood through the ancient borough.

The expansion of Cardiff in the second half of the nineteenth century was phenomenal. Demand for Rhondda's steam coal became insatiable, additional docks being built in 1855, 1864 and 1875 to export the black gold. With prosperity came a rapidly rising population. From 11,400 in 1841, the numbers shot up to 41,400 in 1861; 82,700 in 1881 and 164,300 by 1901. The city of Cardiff was a creation of the Victorian era, and the swelling crowds inevitably brought with them growing thirsts.

Local businessmen and enterprising immigrants were not slow to see and seize the opportunity. Hunt's Directory of 1848 lists six commercial brewers; Slater's of 1850 has eight and Ewen's of 1855 records eleven. By 1863 there were 104 pubs. The frothing glass of Cardiff beer did not stop filling there. By 1880 Butcher's Directory lists sixteen breweries, the number reaching a fine head of eighteen in 1885. This does not include the many English and Irish breweries by now selling into the boom town.

Meanwhile, the Old Brewery continued to change hands like the prize in a pass-the-parcel game. William Williams is last recorded in charge in 1855, and by 1858 the St Mary Street business was being run by Frederick Prosser. An advert of the time says the company were beer and porter brewers, maltsters, hop merchants, wine and spirit merchants and bottlers of ales and stout.

The new owner reflected the rapidly changing face of Cardiff. Frederick Thomas Meredith Prosser was a product of the British Empire; a young, well-travelled merchant, involved in shipping, he had become a mason in 1848, joining a lodge in Calcutta when only twenty-one. He married the daughter of an Indian Army officer, Maria Adams, and their daughter Adeline was actually born at the Old Brewery on 19 September 1859. The bustling new world seemed to be at his feet but, like many before him, an enterprising career was cruelly cut short when he died the following year while on a ship to India.

In 1862 the brewery was taken over by the Thomas family, who were already brewing in Frederick Street. They sold their original Castle Brewery to Henry Anthony, and concentrated the family business at the Old Brewery. It shows the standing of the premises in St Mary Street, behind The Albert pub, that two established Cardiff brewers (Williams and Thomas) had both left their own sites when the opportunity arose to move there. The Old Brewery was not only the oldest brewery in Cardiff; it was also regarded as the best.

John Griffen Thomas and Edward Inkerman Thomas, trading as Thomas Brothers, developed a small chain of 'retail establishments' including two famous Cardiff pubs still standing today, the Golden Cross in Custom House Street and the Old Arcade in Church Street.

Their sister, Frances Elizabeth Thomas, developed a more significant relationship. In 1872 she married Samuel Arthur Brain, a descendant of an eminent Gloucestershire family. His grandfather, William Brain of Kingswood, was one of the leading mining engineers of the early nineteenth century. Samuel Brain had trained as a scientific brewer and had come to Cardiff in 1863, working at and later managing Dowson Brothers' Phoenix Brewery in Working Street, one of the growing number of new breweries in the expanding town.

In 1881 the brewers in Wales received a body blow when the Welsh Sunday Closing Act was passed by Parliament, shutting public houses on the holy day. According to tradition, John Griffen Thomas, who had by then bought out his brother, threw down his pen on hearing the news and exclaimed 'Anarchy!' He resolved to sell the brewery, and no-one was in a better position to buy him out than his brother-in-law.

Samuel Brain was not only an experienced brewer, he also had wealthy relatives. His uncle, Joseph Benjamin Brain of Clifton, near Bristol, was chairman of the West of England Bank and a director of the Bristol Gas Works. The two together bought the Old Brewery in 1882, Joseph Brain providing the bulk of the capital while Samuel Brain ran the business as senior partner. He was a remarkable man.

The Old Brewery, though the leading brewery in Cardiff, was still little more than a three-storey stone building behind The Albert. It occupied only a fraction of the space of the

SUNKEN CELLARS: the remains of an old cask label showing the brewery built in 1887 and the cellars beneath the yard

current brewery. It was certainly much too small for a man of energy and ambition like Samuel Arthur Brain. He determined to build a completely new brewery to meet the demands of the mushrooming population.

He bought land behind the original premises and there, in 1887, at a cost of £50,000, erected a new brewhouse and fermentation block. The towering buildings contained what was then the largest brewing plant in South Wales, including a vast cellar beneath the yard. It was a huge investment for the time. In one bold move he had made Brain's the premier brewer in the region.

Alfred Barnard, on visiting South Wales in 1890 while compiling his weighty work, *Noted Breweries of Great Britain and Ireland*, headed straight for Brain's. It was one Welsh brewery he was determined not to miss. One can sense his eager expectation, as he wrote:

> As the visitor enters beneath the archway, he is confronted by the new brewhouse, which is an elegant structure built on the tower principle, with red and white bricks. It is almost fire-proof, iron, brick and concrete forming the principal elements in its construction, whilst its walls are nearly two feet thick. The iron joists which support the massive floors are as broad as a man's body and, with the numerous metal columns (70 in number) which support them, weigh nearly 50 tons.

The premises were of the latest design, including three steam lifts and elevators for carrying people and materials. Even the three gas-lit cellars were a marvel, floored with vitrified panel bricks and protected from the heavy traffic in the yard above by massive iron girders on iron columns. The central cellar was 300 ft long by 200 ft wide, capable of holding 5,000 casks. In the first cellar a new 400 ft artesian well had been sunk, powered by duplicate pumps. Barnard was captivated.

'The mash ton is one of the handsomest we have ever seen, being constructed of cast-iron, encased in varnished pine, and bound with massive brass hoops.' He thought the adjoining copper-house 'the prettiest and most complete we have visited'. As for the fermentation department: 'This is a delightful room, lighted by 25 windows, and the ventilation is simply perfect.'

Of more importance to Samuel Brain than the 'handsome iron palisading' and 'polished brass rails', which dazzled Barnard, was the vastly increased brewing capacity. The two domed coppers could each hold 150 barrels of beer, while the fermentation room contained six white-cedar vessels each of 150 barrels. It was a huge improvement. The new brewery could produce seven times more beer than the old one.

As sketches of the period show, these new buildings overshadowed the original brewhouse. When Samuel Brain built them he might have expected to close the old plant, but three years later Barnard revealed this was still in operation, such was the demand for Brain's beer. 'Operations are now carried on both in the old and new premises with unceasing vigour, a gyle being brewed every day throughout the year, Sundays alone excepted.'

Barnard found the new size of the business awkward to avoid on his visit. 'As we passed through, the floor was covered with many hundreds of casks that had been recently filled, and we had some difficulty in picking our way between them.' The party did, of course, pause for a drink in the sample room.

DARK SECRET: a poster advertising Brain's Red Dragon dark – the beer on which the success of the brewery was built

Ranged on tiers are casks of every gyle, for future reference and for customers to taste. At the invitation of our courteous guide, we stepped inside and sampled two or three sorts – pale ale, mild beer and stout. The first we found to be a delicious, full-flavoured tonic beverage; the latter a nutritious well-bodied drink, quite equal to the brews of London and Dublin.

But Samuel Brain was not leaving sales of his beers to chance recommendations. Backed by his wealthy uncle, he was buying pubs on a scale that still leaves its mark on the city today. In 1882, when he took over, the Old Brewery had just eleven houses. By 1900 Brain's owned or leased over eighty pubs and, according to one town directory, 'the signboards of the numerous tied-houses owned by the firm are familiar local landmarks'.

When Samuel Brain applied in 1889 to give evidence before the Royal Commission looking into the Welsh Sunday Closing Act of 1881, he stated in his application, 'I am the owner of nearly every licensed house in Bute Street, Cardiff.' The long road, running from the docks to the city centre, was peppered with public houses.

The controversial Sunday Closing Act had also triggered a fresh sector of trade. In 1881 there was only one licensed club in Cardiff; five years later there were 141 as drinkers discovered a way round the dry restrictions. Many of these new premises were supplied by Brain's.

A commercial survey of 1893 says Brain's Brewery controlled 'an enormous local trade'. The original brewhouse had produced little more than 100 barrels a week. With the new plant, 'it is not at all unusual for 1,100 barrels of beer weekly to be turned out', head brewer Mr G.J. Gard revealed in 1890. Six years later the figure had reached 1,300–1,400 barrels a week.

An astute businessman, Samuel Brain had his busy fingers in many related enterprises. He was one of the founders and chief shareholders in the Cardiff Malting Company, registered in 1886, whose huge premises near the docks at East Moors not only supplied many of the local breweries but also companies as far afield as Burton-on-Trent and Ireland. In addition, he was chairman of wine and spirit merchants, Stevens & Sons.

But Samuel Brain was much more than a successful industrialist. Like many other leading brewers, he was active in public life. He was elected to Cardiff town council in 1885 as a representative for Grangetown. He served on the town's school board and many municipal committees and became so prominent in local affairs that in 1899 he was made Mayor of Cardiff.

Two years earlier, he had crowned his developments at the brewery by turning the firm into a limited company. S.A. Brain & Co. Ltd was registered in April 1897, the business being valued at £350,000. The new company remained firmly in family control, the board reading like a Brains' trust. Samuel Brain was chairman and the other directors were his uncle Joseph Brain and Joseph's two sons, Joseph Hugh Brain and William Henry Brain.

Samuel Brain died in 1903, aged about sixty. He had transformed a small town brewery into a major city concern (Cardiff was granted city status in 1905). Even his initials still linger on the lips, as the name of Brain's revered premium bitter, SA.

What followed was the relative calm of stability after a storm of activity. Joseph Brain was chairman for four years followed by his two sons. More houses were gradually added and the sound of dray-horses hooves receded into the past as mechanized transport was introduced. A Yorkshire steam wagon was bought in 1907 and by the First World War the company owned six such vehicles.

Fred Jarvis, who joined Brain's in 1917 as an apprentice, remembers the steamers well. 'I think they did 8 m.p.h. They were coal-powered. The driver was inside by the boiler stoking up and his mate sat outside in all winds and weathers doing the steering with a great big wheel.'

Carrying heavy hogsheads (54-gallon casks) or stacks of crates, they were not easy to control given their laboured steering, suspect braking and solid metal wheels. The driver's view of the road was not helped by the vertical chimney at the front. Not surprisingly, there were frequent accidents despite the slow speeds.

'One wagon going up Warren Hill in Barry slipped back – right into a pawnshop', recalled Fred Jarvis. The driver kept his head, leaped out and asked the distraught pawnbroker, gazing in amazement at the wagon in his shop, 'How much for this lot?'

Many of the draymen preferred genuine horse power and not just because the animals were easier to control. They also could steer themselves. 'I remember one old fellow, Tom Warren. He did local deliveries and he'd get a pint in every pub', said Fred Jarvis. 'At the end of the day he'd be snoozing in his seat and the horses would bring him back to the yard.'

BUSY DRAYS: despite the deep Depression, there's no slackening of activity in Brain's Brewery yard in the 1930s

Like many family firms there was great loyalty among the workforce, some staying all their working life. Fred Jarvis retired in 1971 as chief engineer. The loyalty was reciprocated by the management. Few were dismissed. Fred remembers seeing one of the bosses holding a drunken drayman's head under a gushing cold water tap in the yard to sober him up rather than sack him.

There was no cooling off the call for Brain's beer. Although only built in 1887, the new Old Brewery was straining to keep up with demand by the First World War. So Brain's decided to build another brewery – but this time away from its tight city-centre site.

The company had already started moving some operations out of the Old Brewery in the 1880s when it used additional premises in Womanby Street after taking over Watson's Cambrian Brewery. By 1897 it had a bottling store further away, in the Cardiff suburbs, in Nora Street, Roath.

Before 1914 a larger site between Nora Street and Helen Street was bought, a row of cottages being converted into a new bottling store. And just before the guns began firing in anger, the New Brewery's tower was topped off. The war ended any further building work, and the first brew was not put through until 1919. The stark red-brick building reflects the fact that it was completed in a period of post-war austerity. At the same time the original brewery buildings behind The Albert in St Mary Street were demolished.

The Old Brewery continued to brew Brain's traditional draught beers, notably their popular dark (officially called Red Dragon), while the New Brewery produced beer for the growing bottled trade. Adverts of the 1920s and 1930s show a wide range of bottled brews including Brain's ale, bitter ale, mild ale, strong ale, nut brown, pale ale, IPA, stout, extra stout and home-brewed, available in flagons, pints and half-pints. Extra stout and home-brewed were also sold in nips.

Besides handling its own brews, the New Brewery bottled the beers of other companies sold in Brain's houses like Guinness. These were shipped in bulk to the Nora Street site and conditioned there. Part of the New Brewery is still called the Bass cellar, though Brain's have long since stopped bottling the famous Burton beer.

In 1932 the New Brewery began producing a range of Brain's own soft drinks and mixers, and in 1965 the site also tried a keg beer called Gold Dragon followed by Tudor Light. Both were later replaced by Capital Keg, but none of these processed brews was popular. Drinkers still preferred Brain's traditional draught ales.

Their popularity was so strong in Cardiff that for many years Brains barely bothered to advertise. Apart from the odd corner of a newspaper and their constant fascination with decorating railway bridges, Brains relied on painting their own pubs. The effect could be startling, with huge wall signs proclaiming 'It's BrAIns You Want'. The AI in Brains could be so tall, the letters looked like radio masts.

The AI design is reputed to have been devised by Albert Jones, a glass painter employed by the brewery at the turn of the century. Some of the older houses still carry the blue and white design in their windows. A barmaid, Mrs E. George, claimed to have composed the slogan 'It's Brains You Want' while working at the Canton Hotel just after the First World War.

The words were not always welcome. Temperance campaigners found the slogan hard to swallow, and in the 1930s all drinks advertising was banned from the sides of trams and buses. Many local landmarks disappeared after the Second World War when the planning authorities objected to Brain's wall paintings and hoardings.

The famous family firm gradually built up its tied estate piecemeal, to around 120 houses, the majority in Cardiff itself. The largest single purchase – for around £100,000 – was probably the five pubs belonging to wine merchants Greenwood and Brown in 1956, including the city's best-known 'chop house', the Model Inn in Quay Street.

Brains are remarkable in the high concentration of their houses in one city. Their name is inextricably linked to Cardiff, where their beers have gained a cult following. Their distinctive taste divides the population, as managing director Christopher Brain, great-grandson of founder Joseph Brain, admitted: 'They're an acquired taste. You either love them or you hate them. There's no inbetween with Brains.'

But though the brewery bought some pubs to the west, in the villages and small towns of the Vale of Glamorgan, it owns surprisingly few in the industrial valleys to the north. Here, perhaps fostered by resentment of the prosperous capital city it represented, Brain's ales took a long while to be accepted.

In Cardiff its pubs are legion and legendary. The Old Arcade in Church Street is as familiar to rugby enthusiasts as the National Stadium, while the Golden Cross in Custom House Street is one of the finest examples of Victorian pub architecture in Britain. Its façade is decorated with green and brown glazed tiles, as is the bar which boasts a magnificent mural of Cardiff Castle.

Painful Tail

Nothing illustrates Brain's passion for leaving its mark on every corner of Cardiff, than this moving mural on a brick wall in Bute Street.

First painted in the 1920s, the advertisement shows a poor mutt after a crate of beer has fallen onto its tail from a passing dray. The howling dog is pictured below the punning slogan 'Brain's Honest Ale' (say it quickly).

And nothing illustrates Cardiff's passion for Brain's than the fact that the city council seriously considered saving the threatened mural, and moving it to a new location, as part of Cardiff's contribution to European Heritage Year in 1974.

Sadly the mangy mongrel was finally put down in 1981 when the wall was demolished. But many have not forgotten the

doggone sketch. 'It sticks in your memory because of the look on the dog's face. It was that kind of persistent little terrier – the kind that always runs after you on the street', recalled one admirer.

If you pause by the Salvation Army hostel in Bute Street, some say you can still hear Patch yapping.

It grandly stands at the top of Bute Street, gazing down the straight dive into the docks, an area with its own exotic fascination, as evocatively described by novelist Howard Spring:

The whole place was a warren of seamen's boarding-houses, dubious hotels, ships' chandlers smelling of rope and tarpaulin, shops full of hard flat ships' biscuits, dingy chemists stored with doubtful looking pills, herbs and the works of Aristotle . . . It was a dirty, smelly, rotten and romantic district, an offence and an inspiration, and I loved it.

Tiger Bay once roared with life – and Brain's pubs. None was more famous than the now demolished North and South, where landlady Mrs Bryant acted as unofficial banker for departing sailors, looking after their valuables while they were at sea.

Sailors carried the Brain's name around the world as they developed a taste for the beer. A regular for forty years at one of the few Brain's pubs still left standing in the area, The Packet, recalled the days just after the Second World War:

Everyone drank dark or dark and bitter mixed. There was no lager in those days. Seamen were used to coming in and trying the local brew. It was a man's pub then, with a big coal fire in winter. Men used to come in off the dock, covered in coal dust, for a pint and a warm-up.

Some Brain's pubs have been recorded in literature. The Insole Arms in Harvey Street, Canton, was described in Howard Spring's colourful autobiography, *Heaven Lies About Us*, recalling his early days in Cardiff. As a child, he could only hang around the enticing brass-barred door of the corner pub, which like many in the 1890s was better known by the name of the landlord.

Joe Andrew's public house stood at the intersection of the roads; and in a corner of the wall was the tar-blackened buttress that was Joe Andrew's Stone. The fathers of the street met in Joe Andrew's bright bar; the boys consorted each night at the Stone. For us the Stone was a landmark as geographically important as the Cape or the Horn to more distant wanderers.

Pubs are always landmarks, especially when they stand out from the run-of-the-road buildings around them. Some of Brain's notable houses were designed by acclaimed architects like The Birchgrove and The Westgate, which were both rebuilt in the late 1920s by Sir Percy Thomas. Others had famous licensees. Besides a pack of international rugby players, the most celebrated sporting landlord was probably boxer 'Peerless' Jim Driscoll, the undisputed Flyweight Champion of the World in the early 1900s, who came to run the (now closed) Duke of Edinburgh. His funeral in 1925 was said to have been the largest Cardiff had ever seen. His Lonsdale Belt was bequeathed to his cousin Tom Burns, landlord of the Royal Oak in Roath, where many pictures of 'Peerless' Jim can still be seen.

As Cardiff was transformed into Wales's capital (it gained that long-sought status in 1955) the Old Brewery was left as an affectionately regarded anomaly: a busy industry in a centre of shops and offices. There must have been pressure to leave the inconvenient site – the large drays had to enter the brewery down a narrow side street and then leave through an arch into a busy road, rumbling out in the middle of a set of traffic lights. But the family were not going to be moved from their place in the heart and hearts of a city.

The company was also under pressure of a more welcome nature – the need to produce more beer. Demand for its traditional cask ales shot up, particularly following the formation of CAMRA, the Campaign for Real Ale, in the 1970s. Almost overnight the brewers were ejected from their office to make way for a new stainless steel fermenting vessel. But emergency measures like this were not going to solve its brewing problems.

Fortunately demand for bottled beer was in decline, so some production of its draught beers could be moved to the New Brewery. However, the beer then had to be transported back in bulk to the Old Brewery for racking into casks. This transfer was not without its own peculiar difficulties – or lighter moments.

Once, a twin-compartment tanker, detached from its tractor, was by mistake having its rear compartment pumped out first. Suddenly gravity took over and, pivoting on its central legs, the tanker stood on its nose. A passer-by was heard to remark: 'I never realized road tankers were emptied that way.'

CHANGING FACE: the front view of Brain's Brewery in St Mary Street as it looked in 1934 and the appearance after it had absorbed Paterson's Auction Rooms (Brain's Brewery)

This shuttle service was not a long-term answer, particularly since both breweries were soon brewing to near capacity. In 1975 the decision was taken once more to redevelop the Old Brewery site. It was to prove a long and painful operation, lasting seven years, since brewing and deliveries had to continue throughout the rebuilding work.

First the vehicle service garage was moved into new purpose-built premises half-a-mile away in Collingdon Road. The boilers had to be moved, as they were occupying the site for the new fermenting block, and new bulk malt silos were built. This took until the end of 1977. Work then began in the brewhouse with the

aim of increasing capacity by 50 per cent. Among the new vessels installed were two mash tuns and two coppers. Simultaneously work began across the yard on the new fermenting block, housing six fermenting vessels. Cask racking and washing facilities were also improved.

The fermentation block presented special problems as, being in the city centre, it could not be allowed to look too factory-like. So high-quality glazed finishes were used, giving the appearance of a small office complex. The old brick chimney, built in 1934, was replaced by a thin silver stack.

By late 1979 the essential work of the £2.8 million expansion plan was complete, and the new equipment came on stream, capable of producing an extra 18 million pints a year. But the builders did not disappear. Work then started on up-grading the cellars beneath the yard. This caused extreme problems in the confined brewery, as drays had to steer their way across the yard avoiding the gaping holes open to the sky.

The relief of the staff when the work was finally completed can be gauged from a report in *The Brewer* magazine: 'Eventually, by early 1982 the redevelopment was more or less complete, and a vigorous campaign to evict the contractors – who had taken root – was undertaken.'

It had been an exercise in claustrophobic engineering, as every inch of space had to be used to best effect. Even the flat roof of the fermentation block was designed with a surrounding rail fence so that it could be used for storing casks at peak periods.

The work had also provided the public with a brief glimpse inside Brain's, as buildings were demolished in Caroline Street along the side of the brewery. Passers-by were most struck by the clock, which had been peering down from the top of the 1887 brewhouse since 1926. Once the new fermentation block was built the welcome time check disappeared from public view again.

In 1982 the brewery celebrated a hundred years of cerebral beer with a bottled Centenary Ale, and two years later opened The Old Brewery Shop in St Mary Street at the front of the brewery. It not only sold beers, wines and spirits, but a whole range of Brain's souvenirs from T-shirts and sweaters to tankards and key-rings. 'We have been selling thousands and thousands of items through reception, and with so many inquiries about the possibility of other souvenirs, this has prompted us to open a shop', said manager Alan Morgan. 'We get hundreds of letters from all parts of the world.'

The message 'It's Brains You Want' had spread around the globe.

THE WELSH GIANT

When Samuel Brain built his new brewhouse for £50,000 in 1887, he established Brain's as the premier brewery in South Wales. It was to prove a short-lived supremacy.

Just down the road, another company was fermenting much more fiercely. In little more than a decade from 1883, William Hancock bought eight breweries in South Wales stretching from Newport to Swansea, including five in Cardiff. By the time he died in 1896, he had created a giant, the largest brewing company in Wales.

It was an explosive performance, even in an era of rapid industrial expansion. Unlike Brain's, Hancock's grew by buying brewing companies rather than just buying pubs. The company's arrival shook the South Wales beer glass until the froth flew. But William Hancock's roots were in much gentler soil, the rural calm of Somerset.

His father, also called William Hancock, came from a farming family – with ambitions beyond the local land. One of his brothers was an attorney and another two emigrated to America. William Hancock senior dabbled in various trades. He became a mercer (owning a linen draper's shop) and with his eldest brother Philip founded a bank in Wiveliscombe in 1803. He also became interested in pubs and in his journal of 1805 records: 'May 1st. Began to brew on my own account.'

Two years later, in 1807, he bought land at Golden Hill, Wiveliscombe, and built a larger brewery, which was enlarged again in 1830. On his death in 1845, the business was taken over by his son William. It continued to prosper, becoming not only the major business in the small town but also building a widespread trade. By 1875 Hancock's was described as 'the largest brewery in the West of England'.

Wiveliscombe, some 15 miles inland from Minehead, is only a short hop by ship across the Bristol Channel from South Wales. The coalfields were a booming market the ambitious country brewers could not afford to miss. Kelly's Directory of 1871 records that William John Gaskell and Joseph Gaskell were acting as agents for William Hancock's Brewery of Wiveliscombe, working out of warehouses in West Bute Dock, Cardiff.

SOMERSET ORIGINS: William Hancock learnt the trade at his father's brewery in Wiveliscombe. This company, which remained separate from its Welsh cousin, merged with Arnolds of Taunton in 1927 before being taken over by Ushers of Wiltshire in 1955

HUMBLE BEGINNINGS: William Hancock's first Welsh brewery, the Bute Dock Brewery in Cardiff docks

The Gaskell family were to play a major part in the Welsh fortunes of William Hancock. Their expertise and enterprise were vital. William John Gaskell of Crockherbtown, Cardiff, was familiar with cross-channel trade. In the 1860s he worked in the Cardiff docks for grain, flour and beer merchants John Bowden & Co. of Totnes, Devon, and by the 1870s was acting for himself as a general commission merchant specializing in ale and cider, helped by his son Joseph.

The connection with Hancock's proved mutually beneficial with 'a very considerable business' developing, though the Gaskells did not put all their beer in one barrel. They also sold ale from other breweries. From their Collingdon Road premises they acted as agents for the Burton Brewery, Burton, and for Findlater's stout from Dublin, besides Hancock's.

By 1883 William Hancock decided it was time to take the Welsh trade one step further. Instead of shipping his casks into Cardiff docks he resolved to cut out the sea crossing and brew his beer there. He bought North and Low's Bute Dock Brewery, which had been established in Bute Street since at least 1855. Joseph Gaskell and his brother John became managers of the brewery, which included several pubs, with a share in the profits.

The cross-channel brewing business was not allowed to rest there. Within a year the Anchor Brewery in Mountjoy Road, Newport, was bought from Edwin Hibbard, again with a number of pubs. John Gaskell was appointed manager.

Hancock's was no stranger to Newport, the Gaskells having run an agency in Llanarth Street for Hancock's ales since the early 1870s, though it had been in every sense small beer.

J.A. Green, who began working for the Gaskells in Newport in 1881, recalled: 'We had one horse and trolley and a crank-axle cart. Other than John Gaskell, I comprised the whole of the staff.' Later he became head traveller for Hancock's in Newport, working forty-four years for the firm.

The speciality of the Anchor Brewery in these early days was Hancock's Old Ale. 'Some local brewers took our Old beer, and against this we bought from them fresh beer to supply to our own houses', remembered Mr Green. Newport brewers with whom they traded included William Yorath of the Cambrian Brewery, Baneswell Road, and Elizabeth Evans of the Victoria Brewery in Bridge Street. They also continued selling Burton Brewery beers, Findlater's stout and Devonshire cider.

This Welsh extension was operated as a separate venture and in January 1887, owing to the rapid growth of the trade, William Hancock decided to form a new company, William Hancock & Co. Ltd, to develop the business in South Wales with a share capital of £100,000 and loans of £40,000.

The company was registered in June 1887, with forty-six pubs in Cardiff and thirty-one in Newport. Its telegraphic address was fittingly to be 'United, Cardiff' as it soon combined brewery after brewery in the town.

The Gaskells were well to the fore. Joseph and John were joint managing directors along with one of William Hancock's formidable team of ten sons, Frank, who had been helping to run the Bute Dock Brewery since 1883. Charles Gaskell was secretary.

Frank Hancock was soon making a name for himself in his new country, playing rugby for Cardiff and then Wales in 1884. He became a legend, eventually captaining the national side. In the season 1887–8 he captained Cardiff Rugby Club through an almost invincible year. They were only defeated in their last match by Moseley. Such sporting success cannot have failed to help the new company win friends.

Unlike William Hancock's Wiveliscombe brewery, which continued as a family concern, the Welsh firm was a much more broadly-based business. Also on the board were a number of influential local dignitaries including Valentine Trayes, JP, a director of Hill's Dry Docks; Alderman Patrick Carey of Cardiff; Sir Morgan Morgan; Augustus Filleul of Newport and Robert Bowring of Penarth, a director of the Royal Hotel Company, the Glamorgan Bill Posting Company and the Bank Buildings Company.

The new company had been set up with expansion in mind, and in May 1888 Dowson Brothers Phoenix Brewery in Working Street, Cardiff, was bought, along with fifteen pubs and a maltings in Cowbridge Road, for over £73,000. This was the brewery where Samuel Brain had learnt his trade. The registered offices of the company were soon moved to this more central site from the West Bute Dock warehouse, and Hancock's adopted the phoenix as its symbol. It was an appropriate emblem since the company was rising as fast as the mythical firebird.

In the same year the share capital was increased to £170,000, and Hancock's made its first move west, establishing a branch at Aberavon where the Avon Vale Hotel, two pubs and one off-licence were acquired.

In 1889 the share capital was increased again to £276,250 in order to purchase yet more businesses. The South Wales Brewery, Salisbury Road, Cardiff, was bought in February with twelve pubs from Biggs and Williams for £36,000. In October Hancock's ever-lengthening grasp extended to reach Swansea, buying wine merchants Joseph Hall of Cambrian Place, with fifteen pubs and eight cottages, for £33,706.

In January 1890 Hancock's completed its strategy of running a brewery in each of the three major ports of South Wales, when it bought Ackland and Thomas's West End Brewery in Western Street, Swansea, with one pub, for £6,895. In December 1891 another Swansea brewery was acquired. Thomas Jones's High Street Brewery in Tower Lane, with maltings, thirteen pubs and two cottages cost £10,210.

William Hancock's name now swept the coast, and with each takeover the company gained extra influence and sometimes new board members. Joseph Hall, who became managing director of their Swansea operations and founded Hancock's wine and spirit business, was paid £500 for helping with the purchase of the two Swansea breweries.

The company may have extended even further west. Philip Septimus Dowson, who had joined the board when Dowson Brothers was acquired in 1888, becoming general manager in Cardiff, was appointed the first chairman of Carmarthen United Breweries in 1890. Joseph Gaskell was also involved in the formation of this

RISING FAST: Hancock's early emblem appropriately came to be the legendary firebird, acquired when the company took over the Phoenix Brewery in Cardiff in 1888 (Keith Osborne)

new company, but its poor performance and troubled history put them off taking it over. Dowson, who continued at Carmarthen United, resigned as general manager of Hancock's in 1891.

After this heavy buying spree came the inevitable rationalization of the new business. In Swansea the High Street Brewery was closed in 1894, with all brewing concentrated at the West End Brewery, though malting continued in Tower Lane until 1904. The offices and bottling stores were moved to new premises in Little Wind Street in 1895.

In Cardiff brewing ceased at the South Wales Brewery in 1892, production being transferred to the Phoenix and Bute Dock Breweries. The intention was eventually to brew on one site in the city, but neither of the two premises was suitable. So envious eyes were cast at another company.

The County Brewery was a new concern, having only been established in Penarth Road, Cardiff, in 1889. 'The splendid brewery', gushed a self-penned commercial profile in 1893,

is not only one of the structural ornaments of the town, but is also a very conspicuous object in the view of travellers to Cardiff, either by road, rail or sea.

The building is of the most substantial construction, and measures 103 feet in length, 52 feet in breadth and 63 feet in height. Above it towers a fine octagonal shaft reaching a height of 120 feet. The whole establishment occupies an area of nearly an acre, and is undoubtedly one of the largest and best-organised breweries in the Principality. The plant is a remarkably fine one, and embodies the newest apparatus sanctioned by the experience of the trade.

It was just what Hancock's was looking for.

In 1894 Hancock's bought the County Brewery from the owner, Mr F.S. Lock, together with six pubs. The brewery was unusual at the time in having tried to develop a brand name – Golden Hop – for its light dinner ale. Most beers were just known by the name of the brewers. But it was the potential of the site, situated between the town centre and the docks, which interested William Hancock.

Despite the County Brewery's boasts, the premises were not big enough for his ambitions, only brewing 200 barrels a week. But it had been designed with provision for considerably larger production. The following eighteen months were spent reconstructing the plant and buildings, with new offices, stables, bottling stores and a mineral water factory added. A visitor in 1896 observed:

> Everywhere the latest description of plant and machinery is used, and no effort appears to have been spared to bring the entire business thoroughly up to date. The wide range of ingenious engines and machinery, the excellent order in which all is kept, and the general air of scrupulous cleanliness are, perhaps, the points which are most quickly noted.

The cost of the brewery and the extensive improvements was around £100,000 – and this in a year (1895) when the company was loaning over £50,000 to licensees to secure their business. To meet this huge expenditure the share capital was increased again, being almost doubled to £545,000 by the end of 1896.

The hungry giant also continued absorbing rival businesses. In 1895 Hancock's bought Henry Anthony's Castle Brewery in Frederick Street, Cardiff, chiefly to obtain its trade and twelve pubs. Along with it came memories of a famous court case.

Henry Anthony once sued a publican for not paying for his beer. The landlord's defence was that he had never received delivery. The brewer argued that the claim could be upheld by the drayman who, although unable to read or write, kept a tally on the cellar door in chalk marks. 'Send', said the judge, 'for the man and the door', ordering all witnesses to remain in court. Both were brought and the case was proved.

In 1896 the swelling company added wine merchants Stevens & Sons to its long list of takeovers. The firm in St Mary Street, Cardiff, had once been chaired by Samuel Brain. Now Hancock's took control.

When the revamped premises in Crawshay Street off Penarth Road were opened, the name was changed to simply The Brewery. The other Cardiff breweries were disposed of, the valuable Phoenix site being exchanged for additional land next to the new brewery. Chairman William Hancock was able to tell the AGM in July, 1896:

> In the last report the board expressed their intention of concentrating the whole of the Cardiff business, except the maltings, at the County Brewery, and the concentration has now been effected. The Brewery has been completely equipped; the business of the Phoenix and Castle breweries, as well as that of the bottling department in Working Street, has been successfully transferred to it; a wine and spirit department has been opened, and an aerated waters manufactory has been completed. The result of these developments has already been felt to some extent, and the directors are satisfied that they have all been made to the benefit of the company.

At the same AGM he announced that the directors had experimented with a profit-sharing scheme for employees whereby they would receive 10 per cent of the net profits over and above the trading profits of the previous year. In the trial year this had only amounted to £58, but the aim was to give workers a stake in the company. They were also encouraged to buy shares, special £1 units being introduced for them.

Five months later, in December 1896, William Hancock died at the age of eighty-six. He had created a major rival to Brain's in Cardiff – an internal report shows that even by the early 1890s Hancock's owned a third of the 280–90 tied houses in the town, around ninety-five pubs, and dealt with two-thirds of the free trade. But the shrewd Somerset businessman had also looked beyond the city and carved out a brewing empire running from Newport to Swansea. The company produced nearly 2,000 barrels of beer a week. In thirteen years, ably assisted by the Gaskells, he had established the largest brewers and bottlers in Wales.

But it had not been a constant story of success and backslapping. Hancock's was from 1887 a public company and not all the 360 shareholders were happy with the way the business was run. In July 1893 a shareholders' committee was formed to investigate the firm's affairs.

The complaints were numerous but revolved around the management of the central Cardiff branch:

DOCKLAND HOUSE: an early Hancock's pub, the Pembroke Castle in Louisa Street, Cardiff

* The price of £73,762 paid for the Phoenix Brewery in May 1888, was based on annual profits of £7,463. The actual results obtained in the three following years fell short of this, not exceeding £5,800.
* The South Wales Brewery, bought in February 1889, had been idle 'for some considerable time' yet had not been disposed of – 'the buildings in the meantime having been on the hands of the company, yielding no returns' while costing £150 a year in ground rent.
* The directors had voted the general manager 250 shares for his services in negotiating the purchase of the Phoenix Brewery and 140 shares for helping to buy the South Wales Brewery without gaining the approval of the shareholders.

The shareholders' investigations took the best part of ten months, but in the end largely seemed to vindicate the management, or at least shook them into prompt action:

* The shortfall in profits from the Phoenix Brewery had been 'unfortunate', but the results in the last two years showed 'a distinct improvement'.
* The South Wales Brewery had been closed with 'a view to economy' and after lengthy negotiations, 'which had been set on foot when our investigations commenced', the premises had been sold to the Rhymney Railway for £7,000.
* The directors had acted within their rights in voting shares to the general manager, but the investigating committee wondered about their wisdom in doing so without gaining shareholders' approval.

The knives had obviously been out for Joseph Gaskell, who had taken over as general manager following the retirement of Philip Dowson in May 1891. The committee had been asked to look into his 'want of attention' to the business which it was claimed had caused the company's trade to suffer.

The investigating committee concluded: 'We do not find any justification for the allegation, and in fairness to the general manager we should point out that the company enjoy a larger share of free trade than any other firm of brewers in the town.'

Their report of April 1894 was, however, critical of the running of the Swansea branch, where there had been in 1893 'an astounding loss' of £2,033 in bad debts. Administration charges were also high. 'We see no necessity for the retention of a manager at £500 per annum when such duties can be as effectually discharged by Cardiff.'

They also regretted the drop in trade at Newport and recommended the closure of the brewery there. Although capable of brewing 250 barrels a week, in the five years ending May 1893 the Anchor Brewery had only averaged 123 barrels, under half its capacity.

It was hardly surprising that some shareholders were anxious, for the pace of change and the spate of acquisitions were breath-taking. Yet profits had surged as well, doubling from £16,035 in the company's first year to £32,074 in 1895. 'These figures showed that, if the company had moved rather fast, it had at the same time moved with prudence and safety', managing director Joseph Gaskell reassured an EGM in April 1896, called to raise additional capital. That year profits rose again to £38,080.

Valentine Trayes had taken over the chairmanship of the company following William Hancock's death in 1896, but when he died in 1900 the driving force of the firm, Joseph Gaskell, took the helm. William Hancock's son, Frank, was still on the board, but in November 1896, a month before his father died, the family firm at Wiveliscombe had been incorporated as William Hancock & Sons (Wiveliscombe) Ltd, and Frank returned to Somerset to help his brother Philip run the company as joint managing directors.

With Joseph Gaskell in the chair the company went back on the takeover trail. In 1901 a lease was taken out on the Glamorgan Brewery in Swansea, adjoining Hancock's West End Brewery, providing valuable space for expansion. The freehold with twenty-one cottages was eventually bought in 1920.

In 1902 the Risca Brewery near Newport belonging to Cross and Matthews was bought with twenty-two pubs for £35,500, as the company began to edge up the valleys. Four years later the much-criticized Anchor Brewery in Newport, which even Hancock's admitted was 'somewhat old and obsolete', was closed. A new bottling store was built on the site.

In 1904 the Canton Cross Brewery in Cardiff was bought from a former Hancock's director, John Biggs. He had joined the board following the takeover of his South Wales Brewery in 1889, but resigned two years later when he became involved with Canton Cross.

Joseph Gaskell was pushing hard for business – even to the extent of supplying free beer. In 1904, when filtered flagon ale was first introduced, he decided to give away 3,000 four-flagon cases of ale and stout. A full-page advert appeared in the *Western Mail* giving details of the generous introductory offer. Not surprisingly, the brewery was flooded with postcards. Not wishing to disappoint anyone who had applied within the deadline, he ruled that all these applications must be met. There were over 6,000 postcards and since each flagon contained two pints and there were four flagons per case, this amounted to a gallon of free beer per person – more than 166 barrels given away.

But Hancock's flagon-waving exercise was far from bad business. Agents delivered the gift-packs and then, as bottling manager Tommy Shell later revealed, 'When the agent called a few days after for the empties, he secured in almost every instance an order, and thus in one day, through a daring and generous advertisement, a large business was established.'

In 1909 the Cardiff brewery, which had absorbed the Canton Cross trade and also now brewed for Newport, was extended again following the purchase of neighbouring land and buildings. Part of the new site was for agency bottling and mineral water production (under the Apex brand) and the rest was used for garages to accommodate the growing fleet of motor lorries.

It was a giant plant for its time, capable of producing 2,000 barrels a week. In comparison, Hancock's West End Brewery in Swansea brewed 10,000 barrels – in a year. In the twelve months ending 31 May 1909, the Swansea brewery averaged 208 barrels a week.

The First World War did not stem Hancock's advances, even though the fighting absorbed the chairman. Col. Joseph Gaskell received the CBE for his war services, as a keen leader of the volunteer home-defence forces. He was chairman of the Glamorgan Territorials Association from 1919–22.

Meanwhile, on the beer and business front, David Jarvis's Singleton Brewery in Gam Street, Swansea, was bought in 1917, with the trade transferred to the West End Brewery. A year earlier the Risca Brewery had been closed, with beer for its pubs supplied from Cardiff. Also during the conflict two smaller breweries were bought – the Hanbury Brewery in Caerleon and the Vale of Glamorgan Brewery in Cowbridge.

After the war – when Joseph Gaskell's son Gerald became managing director – expansion was concentrated on Neath. In 1919 the five pubs, one off-licence and bottling stores of Thomas & Sons, Neath, were bought, followed in 1924 by the wholesale bottling and family trade business of H. Stone in Station Street. In the same year John Rees' Abernant Brewery in Cwmgorse to the north was taken over with fourteen pubs.

Three years later in 1927 came a much more prized purchase, when Hancock's captured the Swansea Old Brewery with forty-two pubs. It was the Swansea equivalent of buying Brain's of Cardiff. The acquisition was described in the company's Jubilee handbook in 1937 as 'an important milestone in the firm's history'. 'Among the brewing concerns of South Wales, this one holds a position of distinction', stated a profile of the Swansea Old Brewery in 1893. For once the complimentary words were not influenced by too much beer.

The Swansea Old Brewery was not the first major brewery in the town. In the late eighteenth century a large brewhouse was a prominent feature of the docks, having been

Horse Heroes

*D*ray-horses have always been admired *and* patted by the public. The gentle giants remain popular at shows, even though there are none left working in Welsh breweries today.

Hancock's was famous for its teams of six white horses, which were still used for promotional displays in the 1960s. Almost every Hancock's house seemed to have a picture of them rattling past a pub. But two Hancock's horses in particular once stirred everyone's affection and admiration.

King and Bob were two greys working in harness for Hancock's West End Brewery in Swansea. When the First World War broke out in 1914 they 'joined up', hauling guns for the Royal Field Artillery in France. Danger and death were never far away – 120,000 horses were killed during the conflict. On one occasion their battery was practically wiped out; on another one of the horses was wounded.

But they both survived the horrendous four years to return in 1919 to the streets of Swansea, where they resumed pulling beer for Hancock's. Only this time when they pounded the road they were wearing the 1914–15 Star Ribbons, the Victory Ribbon and the General Service Ribbon. They became celebrities, regularly appearing at hospital carnivals and charity functions, and were still working in tandem in 1929.

established in 1792 by Phillips and Kendall. But it was George Rolls with his Singleton Ales who captured the imagination – and taste – of the public.

He chose his site well, building in the Mysydd Fields, an area noted for its springs. His Swansea Brewery is first mentioned in Pigot's Trade Directory of 1835, and initially it was a humble business, brewing in adapted stables. But demand grew quickly for his beer and by 1850, in partnership with John Hoare, he had established a thriving business in new premises alongside the old ones. Slater's Directory of that year, commenting on the town's industry, noted that some of the breweries were 'large establishments'. In particular 'that of Messrs Rolls and Hoare, the Swansea Brewery, is an extensive concern, and employs a considerable number of hands'.

The brewery's Singleton Well was famous for its depth and the purity and sweetness of its water. During the severe drought of 1887, many householders obtained their supplies from this source. The quality of the water helped the reputation of the beers to sparkle.

In his 1896 publication *In and Around Swansea*, E.E. Rowse noted:

This ale was the local Burton, and was sold to all the publicans round about the country . . . The present company find no diminution in the popularity of the Singleton Ale, inasmuch as a large number of householders and publicans give it the preference over the Burton ales.

The brewery in Singleton Street was run by John Hoare alone from 1851. He was a man of considerable influence and that year managed to persuade the Prince of Wales to donate £250 to the Cambrian Institution for the Deaf and Dumb, which had moved to Swansea. He celebrated his success at a dinner in the Shades Tavern in Salubrious Passage.

The premises were first referred to as the Old Brewery in 1858, by which time the business was trading as Hoare & Co. Like all businesses it needed watching carefully. In 1873

Edward Davies was sentenced to three months imprisonment with hard labour for embezzling money from the company while working for Hoare & Co. as a traveller.

By 1878 there was a new owner, Mr A.W. Cooper, and then in the mid-1880s the Old Brewery appears to have stopped production – until in 1887 a limited liability company, the Swansea Old Brewery Company, was formed. The leading figure and chairman was David Davies, a former Mayor of Cardigan and director of the Cardigan wine and spirit merchants, Davies Brothers. William Davies was secretary. The brothers completely refitted the brewery, employing one of the leading brewery engineers, George Adlam & Sons of Bristol.

The new proprietors were not slow to exploit the reputation of the beers, gaining a prize medal and diploma at the London Exhibition in 1888. They also submitted samples to two eminent analysts whose favourable reports were printed in full on the company's price-lists.

The beers included three grades of mild, the popular drink of the district, plus Crystal Bitter, two qualities of India Pale Ale, a light bitter and three types of stout, 'the Invalid Stout being specially celebrated for its strengthening and nourishing properties'.

In 1896 the Old Brewery was combined with Davies Brothers wine and spirits business and another brewery, Robert George & Son of Pembroke, to form a real mouthful called the Swansea Old Brewery & Davies (Cardigan) Bonded Stores Limited. This new company with fifty pubs was valued at £100,000.

Alderman David Davies was still chairman and managing director, but joining him on the board were his son David, Robert George of the Pembroke Brewery and James Dole, chairman of the Heavitree Brewery of Exeter and a director of Usher's Brewery of Wiltshire. William Davies remained secretary.

The combined venture was never a resounding success, even though it bought a further eight houses in its first full year, and boasted branches in Cardigan, Pembroke, Reynoldstone and Lampeter. 'Great competition' and 'abnormal prices' for pubs squeezed its market, despite the high reputation of its ale.

In 1899 a new company even tried to trade off its name, setting up the Singleton Brewery in Little Gam Street with capital of £3,000. It fooled no one. By December 1900 the receivers were called in. But the attraction of the name never went away. Hancock's itself bought up another attempted Singleton Brewery during the First World War.

Drinkers revered Singleton Ales, particularly 3X, but the share capital of the company was not increased to allow expansion. In 1921 the company's unmanageable title was trimmed back to Swansea Old Brewery Ltd.

When Hancock's bought this respected but drifting company in December 1927, it might have contemplated putting some of its marketing muscle behind the famous Singleton ales. But by now it was investing heavily in the Hancock's name under a new John Bull figurehead introduced in 1926.

In the end it let Singleton ales continue drifting into obscurity. In a time of retrenchment – bottling stores were closed in Neath (1928) and Newport (1931) – Hancock's was content to gain some pubs, remove another competitor and make a profit on the brewery site as the depression deepened.

The Old Brewery in Singleton Street was only a short distance from Hancock's own West End Brewery in Western Street, but it was nearer the town centre – and in demand for a bus depot. In 1934 Hancock's sold the premises to United Welsh Services, who used the site for

PRIZE CAPTURE: the takeover of the Swansea Old Brewery in 1927 was a major advance for Hancock's (Swansea Archive)

a coach station. A famous clock on the brewery, said to be the second oldest public timepiece in Swansea, was moved to the tower of Bishopston Church.

The public transport business was well known to the Gaskells. Gerald Gaskell and his father Joseph had founded South Wales Motors, which Gerald Gaskell remarkably managed to run in tandem with the brewery until pressure of work at Hancock's forced him to sell to the Western Welsh Omnibus Company in 1927.

The pressure was coming from Swansea, and it was the reason behind the purchase of the Swansea Old Brewery. Brewing companies throughout Wales had been shaken by the surprise takeover of the town's other major brewers, Swansea United, in September 1926. The shock was the identity of the purchaser – Truman, Hanbury and Buxton of London and Burton. It was the first major takeover of a Welsh brewery by a leading English company, and triggered Hancock's move for the Old Brewery to safeguard their Swansea business.

Swansea United had been formed in 1890 from an amalgamation of the Orange Street and Glamorgan Breweries. The combined company with fifty-eight pubs was valued at £75,000. The chairman was a former Mayor of Swansea, Alderman Albert Mason, but the prime movers were the joint managing directors Henry Crowhurst and F.D. Mears.

Brewing was concentrated at Crowhurst's Orange Street Brewery, which was completely refitted in 1892. The Glamorgan Brewery in Little Madoc Street was used for the production of non-intoxicant hop bitters, trading under the name of the South Wales Hop Bitter Ale Company. United also dealt in wines and spirits with bonded stores in Little Wind Street and

Oystermouth Road, manufactured mineral waters and acted as agents for some of the Burton breweries.

It was an ambitious venture, employing six travellers to sell the beer and several agents. Its symbol was the Kentish horse Invicta, often seen on hop pockets. In the first year eleven pubs were bought. United even tried to follow Hancock's policy of trading throughout South Wales, opening an office in Hancock's Cardiff backyard, in Plantagenet Street, with bottling stores in Union Street.

Ten years after United's inception, annual production was up to over 30,000 barrels from an initial output of 18,400. Albert Mason was also happy to report at the tenth AGM in 1900 that sales of wines and spirits and mineral waters were 'extremely satisfactory', all showing steady rises. There were setbacks as well. The South Wales Hop Bitter Company failed to take off, the Glamorgan Brewery being leased to Hancock's in 1901, and its attempt to break into the Cardiff market only lasted until 1906. But by 1910 it was able to report that the company had more than doubled its number of licensed properties since it began. And by 1920 United was showing annual trading profits of over £25,000. Every year since its formation it had paid a dividend to shareholders. For 1919 it was 15 per cent. This was in contrast to the Swansea Old Brewery, which failed to declare a dividend some years and when they did fell into arrears on payment.

Thus when Swansea United was seized by Truman's in 1926, it came as a shock. Some feared the fate of the Invicta brewers would prove to be a stalking horse. Other English brewers would follow, muscling their way into the Principality through takeover.

That it did not happen at this time was partly down to Truman's unfortunate experiences. It found the takeover left it with a parcel of pubs out on a limb, which proved difficult to manage from London. Other major brewers preferred to continue selling their beers through agents and agreements with local breweries, though some must have thrown greedy glances at Hancock's.

In 1887, when the company was formed, there were about forty employees. Forty years later the workforce was well over a thousand. In 1887 the share and loan capital of William Hancock was £140,000; in 1925 it had been boosted again to £945,000, with a further £170,000 in reserve. The profits that year topped £70,000. In the beginning marketing had meant a man with a cart and a loud voice. Now there were electric slogans flashing over busy Queen Street in Cardiff.

In 1926 the company ran a fleet of eighty-three motorized drays to speed their beer throughout South Wales. One in particular caught everyone's interest, as it drove daily between Cardiff and Newport or Swansea. The huge tanker on its back, holding five tons of bulk beer for bottling, was in the eye-catching shape of a giant bottle of Hancock's Strong Ale.

But one thing had not changed over the four decades – Joseph Gaskell was still in charge. Ill-health finally forced his resignation as chairman in 1929, Philip Hancock briefly taking over. Remarkably two of the other founding directors, his brother John Gaskell and Robert Bowring, also remained on the board.

Joseph's son, Gerald Gaskell, became managing director in 1925, the same year he was elected to Cardiff City Council, and was made chairman on Philip Hancock's death in 1933. He presided over a period of fresh enterprise, ably assisted by the bubbling bottling manager Tommy Shell.

LOT OF BOTTLE: the famous tanker lorry which trunked five tons of beer at a time between Hancock's breweries

Shell, who became a director in 1934, took a completely new approach to selling beer. He was fascinated by innovations, regularly visiting marketing conventions. He was the one who switched the company on to electric signs, turning the sides of some prominent houses into mini Piccadilly Circuses. The Salutation Hotel on the main Cardiff Road in Newport became famous for its huge neon display advertising Hancock's Amber Ale with a glowing glass.

At the Pure Food Exhibition in Cardiff in 1926 he mounted a colourful stand of Hancock's beers round a large 'ever-flowing flagon' of oatmeal stout that, through technical tricks, poured forever. Always impressed by gadgets, he bought a sales figure which would not be acceptable today – a 4 ft 9 in model of a black page boy, which could move its head, arms, eyes and mouth. Rastus, as 'the boy' was called, was used in hotel lobbies and at exhibitions to distribute leaflets.

Shell regaled drinkers with odd verses:

> Hancock's Stout is here no doubt,
> If not in wood, in bottle,
> And sparkling ale which ne'er did fail
> To quench a thirsty throttle.

And even stranger questions:

> If it takes a double-breasted earwig four seconds to swim through a hogshead of Sandeman's Port, how long will it take a six-foot man, wearing boxing gloves, to pick a blue-nosed wasp out of a flagon of Hancock's Strong Ale, provided there is a rebate of 1½d. per dozen bottles?

His humorous posters became legendary. One striking painting showed a sailor at the seaside knocking back a bottle of Hancock's beer, held up to his face, with a little bucket-and-spade girl asking 'Please may I have a look through your telescope?'

SELLING WITH A SMILE: bottling manager Tommy Shell realized the power of humour in marketing beer with a series of classic comic posters in the mid-1920s

Shell even managed to turn adversity into good publicity. In 1927, when Hancock's was ordered by Cowbridge Magistrates Court to remove its roadside sign pointing to the Bush Inn at St Hilary, following a successful prosecution by Glamorgan County Council for 'disfiguring the beauty of surrounding land', a tongue-in-cheek prize competition was run in the *Western Mail* to find an acceptable sign.

Entries piled in from professional sign-writers to a six-year-old girl. 'Even the amateur futurist has tried his hand. His effort shows a man and woman at table framed in a cubist nightmare', said the judge. 'If such a sign was placed in the verdant quietudes of the Vale of Glamorgan one verily believes that the wrens and the robins would amaze the rustics by chirping in jazz syncopations.' Another ingenious entry sketched in a hand over a weathervane as a compliment to Hancock.

The winner was startlingly simple – a neatly trimmed and shaped bush carrying a lettered sign. Whether this shrubbery totem pole was ever built, or grown, is unknown.

There was a serious side to this smiling approach. The need to sell Hancock's name and beers. In 1926 the old phoenix emblem was crated away and replaced by a proud John Bull figure, designed by a member of the Gaskell family and drawn up by the *Western Mail* cartoonist, Leslie Illingworth, who had judged the inn sign competition.

New beers were introduced – with a fanfare rarely seen before in Wales. In 1928 a fresh brand, Hancock's Nut Brown Ale, was launched at Cardiff Races on Easter Monday, when thousands of bottles were thrust at the eager racegoers.

Hancock's, like all brewers, needed to sell hard. Annual profits had already slipped by nearly £9,000 to £61,451 in 1926 owing to the General Strike and 'depressed trade conditions in the area'. Worse was to come in the deeper Depression after the Wall Street Crash. In 1929 Hancock's failed for the first time to pay its shareholders a dividend. In 1930 it managed 5 per cent, but then for six years from 1931 to 1936 there were no dividends again. Trade was grim.

With the Depression particularly acute in the coalfields and iron and steelworks of South Wales, Hancock's was one of the few Welsh breweries large enough to burst out of the bleak local trade. It started to push its beers into Herefordshire and the west of England. It also had other cards to play.

As well as being Wales's biggest brewers, it was the country's largest wine and spirits merchant – The Wine Cellars of Wales – supplying not only its own 450 pubs and off-licences but also other wholesale and retail merchants.

The brewery had been involved in wines and spirits since taking over merchants James Hall of Swansea in 1889, followed by Stevens & Sons of Cardiff eight years later. But it was not until 1932 that the company took a long, hard look at this side of its business. As a result the department was completely reorganized and in five years sales leapt four times despite the tough trading conditions. In 1932 the bottling cellars only held twenty-nine 'standing pieces' (great storage casks); by 1937 there were sixty-six. The number of items stocked more than doubled from 809 to 2,011 over the same period. The brewery was the sole agent in South Wales for many market leaders like Pedro Domecq and Gonzalez Byass sherries and ports. Its slogan was 'You can get it at Hancock's'. The range of whiskies included its own brands Duchess and Edinbro' Cream.

Pride pours off the page of a written description of the department in 1937:

The rule in the trade seems to be that the dingier the cellar, the better, and it is quite usual to see the processes of bottling and blending being carried on in surroundings of impenetrable gloom and traditional mustiness, where even the cobwebs seem hallowed by age.

The bottling rooms at Hancock's are in sharp contrast to such a picture, for they are light, lofty, well-ventilated rooms of nearly 3,000 square feet in extent, equipped with the latest types of filters, bottling machines and other plant. Around the walls stand the wine bins, capable of holding about 3,000 dozen bottles, while placed like massive sentries before them are the sixty-six great 'standing pieces', painted in black and white, from which the wines and spirits are drawn off into the bottles.

A walk through these stores must have been a heady experience. There was the Cocktail Room, the Liqueur Room, the Champagne Room and the Tonic Wine Room. Everything was there from homely cowslip and elderberry to vintage brandy. There was even a special room for kosher wines kept for Jewish customers under the supervision of a rabbi. Besides the bottling rooms, there were extensive duty-paid stores in Crawshay Street, Cardiff, and Little Wind Street, Swansea, with a floor area of 12,000 sq. ft. Not to mention tobacco stores selling half-a-million cigarettes alone every week.

Hancock's business did not stop there. In addition to being the leading brewers and wine merchants in Wales, it savoured another much more unusual ambition – to become the leading caterers in Britain.

Most brewers at this time fought shy of food, reluctant to even concede to their customers a pie with their pint at the pub. Hancock's, however, dived into dinners with relish – but what was on its menu was not a few bar snacks.

Its catering career began thanks to the Gaskell family's close connections with the army. Both Joseph and Gerald always liked to be known as Col. Gaskell, Gerald having served in the Royal Artillery in France, Egypt and Palestine during the First World War. Both were prominent in the Territorials after 1918, and their first catering contracts came from the army in 1925.

These military manoeuvres provided valuable experience. Gradually Hancock's built a reputation for efficiency on a grand scale, for being able to provide extensive meals for hundreds of people often in awkward surroundings. Organization – military organization – was the key.

That catering, especially under the difficult conditions imposed by out-of-door functions, is an art in itself, will not be disputed by anyone who has ever tried to induce a kettle to boil over a picnic fire, wondering at the same time how to dare break the news that he has forgotten the butter', explained Hancock's catering department. 'To forget several hundredweight of it when catering for a luncheon for 2,000 would be a catastrophe.'

Hancock's lived up to its slogan 'We cater anywhere.' Not only did the brewery handle engagements throughout Wales, but also across England. By the mid-1930s annual events included the Richmond Royal Horse Show, the Bath and West Agricultural Show, Kent and Suffolk county shows, international sheepdog trials, golf championship meetings and the Eton v. Winchester cricket match, besides many army camps, hunt balls and club dances and dinners.

Hancock's was official caterer to the Welsh Golfing Union, Swansea Guildhall, Cardiff Race Club and Malvern Winter Gardens. It organized functions for councils in Cardiff, Wolverhampton and Brighton. The latter two were because the brewery had taken over the catering business of Smith and Berry in Dudley and the restaurant, Booths of Brighton. Hancock's also boasted the popular Connaught Rooms in Cardiff, where 750 guests could be entertained at one time.

The scale of these operations was immense. In 1937 Hancock's handled the coronation celebrations for Fulham Borough Council, which involved, just for starters, feeding 25,000 London schoolchildren. Apart from food, Hancock's had to be able to provide everything else from tables and chairs to erecting vast marquees and laying dance floors on lawns. The operation had come a long way from army-cooking tents.

The company was also building up a chain of residential hotels like the Alexandra in Cardiff and the Wyndham in Bridgend, besides some on the coast such as the Barry Hotel near Barry Island, the Pier Hotel at Mumbles on the Gower peninsula and the Jersey Beach at Aberavon where it expected a Welsh Blackpool to be created.

These extra strings to its bow helped Hancock's ride out the Depression. Unlike many breweries it was not bent completely over a barrel. And its catering arm gave it a chance to reach out beyond the grim conditions in the Welsh coalfields and iron and steel industries.

This did not mean that it neglected its bread and bitter trade. Hancock's continued to pick up pubs, particularly in the valleys. It bought one parcel of a dozen houses from George's when the large Bristol brewers decided to pull out of South Wales.

Hancock's crowned its push north in 1936 when it bought one of the largest and oldest breweries in Merthyr Tydfil, Giles and Harrap's Merthyr Brewery in Brecon Road with sixty-two pubs. This was the year the iron and steel town's independent brewing tradition came to a sudden end as Andrew Buchan's Rhymney Brewery also snapped up David Williams' imposing Taff Vale Brewery.

Hancock's and Rhymney's hobby-horse brewery were jockeying for position as trade at last began to pick up. Rhymney also bought Crosswells Brewery in Cardiff in 1936. The *Daily Express* commented: 'All the South Wales breweries are doing better after several lean years. Those in control believe that recovery can be hastened by the formation of more powerful units.' Hancock's was already a formidable force. Buchan's was pushing hard.

To meet the challenges ahead two new floors were added to the fermenting block at the Cardiff brewery, with new aluminium vessels replacing older wooden ones.

The rivalry came to a halt with the outbreak of war. Production at the brewery almost came to a standstill as well. Gerald Gaskell's son Joseph, who had started working for the company in 1935, was as keen on the Territorial Army as his father and grandfather. His enthusiasm for recruiting fellow employees meant that there was a mini manpower crisis at the brewery when the Territorials were called up in 1939.

His father Gerald was just as active on the home front, commanding the Cardiff Home Guard and becoming High Sheriff of Glamorgan in 1941. He was also prominent in wider British brewing circles, being chairman of the Brewers' Society's wartime publicity committee. On demobilization in 1946 Joseph Gaskell was appointed assistant managing director and then joint managing director three years later.

During the war Hancock's catering expertise was harnessed to the war effort, the department running vital industrial canteens. After victory was declared, the catering corps continued to advance taking over Comleys of Porthcawl in 1947 and the business of W. Osmond & Sons of Salisbury in 1950.

The knife and fork brigade had led Hancock's into unfamiliar drinking territory. By 1939 it already owned odd pubs well away from South Wales, like The Grapes in Leominster and the Lamp Tavern in Dudley, near Wolverhampton. After the war it tried to exploit these distant markets for its beer.

Between 1946 and 1950 Hancock's leased the redundant premises of the Queen's Cross Brewery behind the Lamp Tavern in Dudley as its West Midlands depot. But the Cardiff ales never captured the tastes of the Black Country drinkers. The depot was always overshadowed by Julia Hanson's brewery opposite, and the brief venture into the Midlands was abandoned.

Of much more significance for the future was Hancock's ever-growing link with Bass. The company had strong trading connections with another famous Burton brewer, Worthington's, reaching back to the last century. In 1896 Joseph Gaskell even claimed that the quality of its beer was a prime reason for its excellent profits. The *Brewers' Journal* reported:

These results were due, not to any cutting of prices, but principally to two factors: the loyal efforts of their staff, of whose help he could not speak too highly, and the excellent quality of the India Pale Ale supplied them by Messrs Worthington. Their present stock of this ale was really of most excellent quality, and so long as Messrs Worthington maintained this quality they might reasonably expect that their sales of East India Pale Ale in bottle would be perfectly satisfactory.

In 1927 Worthington merged with Bass, Ratcliff and Gretton.

After the Second World War the relationship became cosier – Bass making Gerald Gaskell chairman of one of its subsidiary breweries in London, with his son Joseph as a director. The company was the Wenlock Brewery of Shoreditch, in which Bass had held a controlling interest since 1922 before fully taking over in 1954.

In 1959 Gerald Gaskell, who had retired from full-time business in 1955, died. His son Joseph was made chairman of Hancock's and in the same year cemented the relationship with Bass by signing a long-term trading agreement. In 1961 he became a director of Bass. A Bass director, W.P. Manners, was already on the board of Hancock's, the Burton brewers having taken £50,000 worth of shares in 1958 to replace a loan.

This close connection protected Hancock's from takeover during the sixties, a decade when independent brewers fell like pins in a pub skittle alley. Except, of course, there was no protection from takeover by Bass.

Hancock's was by now a prize target, with 440 licensed premises and two breweries. Modernization continued at the Cardiff plant where capacity was increased to 4,500 barrels a week. In 1960 it reported a healthy profit of £350,955. In the same year it made its last takeover bid.

David Roberts & Sons of Aberystwyth was West Wales's largest (and almost only) brewer. Founded in the late 1850s by maltster David Roberts in Trefechan, the family firm was registered in 1897 as a private company with capital of £20,000 restricted to the Roberts family.

Though the company ruled central West Wales, building up an estate of a hundred plus pubs, it was never a large firm, many of its houses being scattered through country villages. The brewhouse had a capacity of no more than 400 barrels a week, being best known for its Crown Ale and Rocket Stout. Business was often a struggle.

In 1890 the family firm had tried to break out of its immediate trading area by linking up with Thomas Issard's Crown Brewery of Newtown and wine merchants Mytton & Galloway of Welshpool to form the Montgomeryshire Brewery Company – but this joint venture was a failure and the arrangement rapidly collapsed. Success proved just as hard to find in later years.

The first public company, formed in 1935 with a capital of £152,000, went into liquidation in 1941. A new company was created and 45,000 shares were issued in 1949 to finance expansion to the east. In 1950 Roberts bought up the small family firm of Facey & Son of Abergavenny with twelve pubs, following the death of the owner Frank Facey in 1949.

The Abergavenny brewery was closed, being used as a store, but the move left Roberts with an awkward cross-country delivery route. Nearly 200,000 more shares were offered to existing holders in 1951, but soon the company was struggling to pay a dividend. None was paid to ordinary shareholders in 1952 and for five years from 1954 to 1958.

In 1957 Roberts tried to manoeuvre itself out of its difficulties by a daring liaison with two of Britain's largest brewers. It joined with Hope and Anchor Breweries of Sheffield (brewers of Jubilee Stout and Carling Lager) and Ind Coope's wine and spirits subsidiary Grants of Burton-on-Trent to form a new company called Facey Ltd.

Run from premises called Jubilee House in Llanfoist, near Abergavenny, its aim was to sell a range of beers, wines and spirits in the area. But this plan was soon scuppered when Hope and Anchor became part of United Breweries and turned to marketing its popular brands in South Wales through another United member, Webbs of Aberbeeg.

Early in 1960 Hancock's bid £680,000 for Roberts. Chairman T.G. Boardman quickly recommended acceptance. The takeover went through and Roberts brewery soon closed. Hancock's paid for the acquisition and the spending required to improve the 117 pubs by another share issue, bringing the capital of the company up to £1.8 million.

Hancock's had for many years maintained a presence in Aberystwyth through its ownership of the White Horse Hotel. Now its position in West Wales was considerably strengthened. The takeover also helped boost profits by over £150,000 to £503,113 in 1961, with a dividend to shareholders of 17½ per cent.

Chairman Joseph Gaskell said the takeover had 'proved to be worthwhile', but he was more enthusiastic about another development – the introduction of Hancock's own keg beer called Barleybrite which had started to appear on Welsh bars in 1960. Sales were 'outstanding', though the cost of new plant to produce it was also high. A tank beer, Superex, was added three years later, along with a keg version of its popular PA called Allbright.

There was also heavy spending (over £350,000) on a new bottling plant at the brewery. When it was officially opened by the Lord Mayor of Cardiff in 1963, it was said to be the most modern in Europe, capable of filling 175,000 bottles a day. The 18,000 sq. ft hall had taken two years to build, doubling the amount of beer that could be handled. The older bottling lines at Swansea and Aberystwyth were then closed. Hancock's was by now sole agent in South Wales for Bass, Guinness and Carlsberg.

The mayor commented at the opening that the brewery was both 'a family affair' through the Gaskells and 'a city affair' as Joseph Gaskell and secretary Ronald Tucker were City Councillors. It was no idle comment but a statement of solidarity, as the 'family affair' was looking increasingly under threat.

Hancock's close relationship with Bass had been jolted when the Burton brewers had merged with Mitchells & Butlers of Birmingham in July 1961 to form Bass, Mitchells & Butlers. To outsiders it looked like Bass had taken over another company; the reality was that the more aggressive M&B management was now in charge under chief executive Alan Walker. Hancock's soon made a trading arrangement with M&B, exchanging its premium bottled beer Five Five for M&B Export, but the old cosy friendship was never the same. This was the period when massive British brewing combines were emerging, notably Ind Coope, Ansells and Tetley-Walker ganging up to form Allied Breweries in March 1961, with an estate of 9,500 pubs. Everyone was becoming nervous.

Takeover talk about Hancock's took off with a vengeance in April 1962 when its shares jumped in value from 9s. 6d. to 12s. 6d. in one day on the Cardiff Stock Exchange. Someone was buying heavily. The pressure was on and Wales's biggest brewer was feeling the strain.

Beer sales were still rising by some 4 per cent a year and profits were good (£611,894 in 1963) but after tax that profits figure was cut to £290,000 – well short of Hancock's heavy expenditure of around £650,000. On 31 March 1963 the company's bank overdraft was £791,622. Joseph Gaskell complained at the AGM that year:

> Once again costs continue to rise, especially on repairs to our houses, and I cannot see any hope that this situation will be reversed except by doing less work on our houses, which would be quite the wrong policy in these days of intense competition.

The redevelopment of many town centres was causing major problems, as Hancock's needed to replace demolished old houses with expensive new ones. The brewery was building four to six a year, sometimes in conjunction with Bass in order to keep down the costs. 'The question of capital expenditure will have to be considered by the board in the autumn and the method of dealing with the problem settled after discussion with our financial advisers', the chairman concluded, adding hopefully, 'What we really need now is a fine summer.'

Even if the sun did not shine the heat remained on. The share price rose by 30s. in the six months from June, and by 12s. 6d. in one week in December. Joseph Gaskell again had to deny 'persistent rumours' of a takeover bid, blaming financial pundits who had claimed that on the prices offered in recent bids, Hancock's was undervalued. 'There is a sort of artificial demand for our shares', he said.

It was a question of running hard to stay ahead of the wolf pack. In 1964 Hancock's recorded record profits of £691,000, up 13 per cent. The dividend was 22½ per cent, enough to keep the shareholders happy. The bank overdraft was £1,085,496. The company raised £900,000 through a debenture issue. 'The idea is clearly to cut the group's bank borrowing, which must be proving expensive', commented the *Western Mail*.

At least by the end of the year, six years of costly development at the Cardiff brewery were coming to an end. It had not been just the new plant and buildings that had been a drain on financial resources, the ground they were built on had also proved expensive. One-and-a-half acres had been reclaimed from the River Taff; it had taken thousands of tons of rock and 250 yards of steel sheet piling to create a new river bank.

In 1965 profits were up again to £761,401 but the takeover rumours would not go away – and the pressure increased when Whitbread swooped to buy Rhymney Breweries in January 1966. Discussions were held with Bass, Mitchells & Butlers in October 1966 when it was revealed that BM&B already owned a quarter of Hancock's, the share stake having being bought by M&B before its merger with Bass.

Bass and Hancock's had an old gentlemen's handshake agreement that Bass would not bid for the Cardiff company without the consent of the Gaskell family. Joseph Gaskell hesitated, Bass waited – and Charrington United stepped in and snapped up shares on the market. Now Bass seemed nervous. It had previously lost a long-standing ally when London brewers Watney-Mann had moved in behind its back to seize the leading Manchester brewers Wilsons in 1960. And in 1967 it was involved in another battle with Watneys to gain control of Bents of Liverpool. This time, after a five-week struggle, the Burton brewers emerged victorious.

On the other hand, perhaps Bass was not too worried about the fate of Hancock's after all.

Chairman Alan Walker had a crucial card up his sleeve. He was negotiating a merger with Charrington United and in August 1967 these talks resulted in the creation of Britain's largest brewing conglomerate, Bass Charrington, with 11,000 pubs. Any Hancock's shares Charrington had bought fell into Bass's bulging back pocket.

Joseph Gaskell had already begun to make appropriate noises in 1966 when he said the future would be difficult 'especially for smaller companies which going it alone cannot achieve great economies of scale'. In Wales Hancock's might be a giant but in Britain, in the new order of massive combines, it was a minnow.

In 1967 profits were up again by over 12 per cent to £820,264, though it was hit by the new corporation tax. More causes for satisfaction were that assets exceeded liabilities by £387,630 and the bank overdraft had been reduced by £340,110. Hancock's John Bull figure was in good shape – but that only made the company more attractive.

In September the price of its shares leaped once more by 11s. 6d. to 122s. 6d. A bid was again denied by Bass – but it was on its way. In February 1968 Hancock's directors recommended their shareholders accept a £7.7 million offer from Bass Charrington.

Joseph Gaskell had bowed to the inevitable and was made chairman of the new company of Welsh Brewers, which combined Hancock's 502 pubs and 33 off-licences with Bass

STIRRING SIGHT: skimming yeast in a fermenting vessel at Hancock's Brewery in Swansea. The plant closed in 1969, bringing to an end the town's long brewing history

Charrington's existing interests in South Wales, notably Webbs of Aberbeeg and Fernvale Brewery in the Rhondda, to provide a total estate of some 750 houses. 'Our customers will not lose their favourite brew but will be offered a wider variety of products', pledged Joseph Gaskell. 'And Hancock's devotees will be able to obtain their drink in many more public houses.'

Rationalization inevitably followed. Immediate casualties were Hancock's old West End Brewery in Swansea, which closed in 1969, and the Fernvale Brewery in 1970. Beers like Barleybrite and Five Five vanished. Hancock's celebrated catering arm was sold off and some depots disappeared including ones in Newport and Merthyr Tydfil.

Wales's brewing giants had been swallowed by Britain's brewing giants. The company that had taken over more breweries in Wales than anyone else had finally been taken over itself.

HANCOCK'S TAKEOVER TRAIL

1883: North & Low, Bute Dock Brewery, Cardiff.
1884: Edwin Hibbard, Anchor Brewery, Newport.
1888: Dowson Brothers, Phoenix Brewery, Cardiff.
1889: Biggs & Williams, South Wales Brewery, Cardiff.
1889: Joseph Hall, wine merchants, Swansea.
1890: Ackland & Thomas, West End Brewery, Swansea.
1891: Thomas Jones, High Street Brewery, Swansea.
1894: F.S. Lock, County Brewery, Cardiff.
1895: Henry Anthony, Castle Brewery, Cardiff.
1896: Stevens & Sons, wine merchants, Cardiff.
1901: Glamorgan Brewery, Swansea.
1902: Cross & Matthews, Risca Brewery, near Newport.
1904: John Biggs, Canton Cross Brewery, Cardiff.
1914: Hanbury Brewery, Caerleon.
1915: Vale of Glamorgan Brewery, Cowbridge.
1917: David Jarvis, Singleton Brewery, Swansea.
1924: John Rees, Abernant Brewery, Cwmgorse.
1927: Swansea Old Brewery, Swansea.
1936: Giles & Harrap, Merthyr Brewery, Merthyr Tydfil.
1960: David Roberts & Sons, The Brewery, Aberystwyth.

Castles and Christmas Puddings

If Cardiff and Swansea were known for their large breweries, that other prominent port on the South Wales coast, Newport, is probably best remembered for its eccentric brewing businesses. One in particular stands out as a monument to Victorian values. Wales is famous for its many castles, but only in Newport was a commercial brewery established in the ruins.

Newport Castle had been built alongside the River Usk in the fourteenth century, commanding the river crossing. By the mid-eighteenth century it was in a dilapidated condition, with trees growing out of the walls. In the early years of the next century the bailey was in use as a tanyard. Then in 1828 barrels and brewing vessels rolled into what was left of the main fortifications.

FORTIFIED ALE: a sketch of Newport Castle around 1830, already steaming away as a brewery (Cadw), with a later trade card issued by Edward and John Allfrey (Welsh Folk Museum)

To men of business it made perfect sense. The neglected castle had three large towers ideal for converting into a tower brewery. The thick walls offered cool insulation; the cellars were perfect for storage and the well provided good brewing water. Its site next to the river also meant transport was no problem. This use of ancient monuments for industrial activities was far from unusual at the time. Even religious sites were not sacred. Copper smelting was carried out within the walls of Neath Abbey in the 1790s.

The first rude intruders into what had once been a noble residence were Ramsbottom and Allfrey's. By 1835 the two brothers Edward and John Allfrey were running the Castle Brewery on their own. If any of the town's civic leaders disapproved of their choice of site, it didn't show. Castle Brewery beer was welcomed at the most prestigious occasions.

When the Mayor of Newport, John Owen, cut the first sod to start work on the town's dock in December 1835, church bells were rung and guns fired before a cheering crowd of several hundred at Pillgwenlly. Then the contents of several barrels of Castle Brewery XXX were liberally distributed among the navvies 'who seemed to be on particularly good terms with the beverage of that establishment', noted an observer.

It proved useful business for the Allfreys as work lasted on the dock for six years. The brothers are described as ale and porter brewers, maltsters and hop merchants. The prominent brewhouse prospered in the rapidly expanding town. Another member of the family, Francis Allfrey, was in charge by 1871, but within a few years a former manager Alfred Blake had taken over. Then came Messrs Searle and Herring with grander ambitions.

Richard Searle and Alfred Herring bought the Castle Brewery around 1881 and were soon planning to turn it into a limited company. A prospectus of 1889 stated that the annual profit for 1888 was £6,576. The company was registered with a nominal capital of £80,000, the fortified brewery and fifty-seven tied houses being bought for £65,000. Local JP Arthur Evans was chairman, and Alfred Herring and Richard Searle junior were joint managing directors. The venture seems to have been a success. In the first two years dividends of 9 and 10 per cent were paid. In the first three years twelve more pubs were bought.

A commercial profile of 1893 described the business:

> The remains of the old castle comprise a block of building with one central and two wing towers facing the River Usk and close to Newport Bridge. To these old relics new premises have been added, including offices, stables and other departments. The castle (apart from the southern tower by the bridge) is used for brewing purposes, and has been well arranged and fitted up with every known appliance and machinery . . . upon the most modern principles.

The plant was a fair size 25-quarter one, with a 20 horse-power steam engine supplied by Adlams of Bristol. 'The subterranean dungeons have been converted into very serviceable cellars and stores, and are now filled with extensive supplies of October and other special brewings, amounting in the aggregate to as much as 70,000 gallons.' The writer claimed that the 'extensive trade' reached as far as Herefordshire, Somerset and Devon. Besides light dinner ale, mild, stout and pale ale, the brewery also produced 'vatted ale' and 'castle underground', otherwise known as XXXX.

The company employed some thirty men. For those who were required to live on the premises, it proved an unnerving experience, as Herbert Jones later recalled. 'I was three at the time and remember quite a lot about the interior of the brewery with its long cellar containing huge vats of beer.' His father was cellar foreman. The family did not stay long. 'My mother did not like living in the castle, mainly because of the noise of the brewery horses at night. They were stabled near the house.' Because of this his father found another job.

Others were equally uneasy about the location of the brewery. Even the writer of the company profile thought it odd, perhaps a sign of retribution:

> When in the long past . . . the founders built the fortified castle of Newport, it could not by any process have entered into their heads that the nineteenth century would see their architectural achievements utilised for 'ye brewynge of ale'. Yet such is what has come to pass, and the whirligig of time has had, as Shakespeare says, its revenges.

Some visitors were angered and depressed by what they saw. Wirt Sikes, the American consul to Cardiff, recounted his colourful reaction to this 'wretched and grimy ruin standing in melancholy dejection' in his book *Rambles and Studies in Old South Wales* (1881):

> Of all the old castles in Wales, perhaps this is the most mournful to look upon, so fallen is it from its grand estate. A ruined castle is seldom debased to such plebeian uses as those which have befallen Newport. It is now occupied by certain brewers and other unknightly varlets; but there was a day when this pile was the home of kings; and where now ferments the democratic ale and rings the rattle of barrels being hooped, once flowed the ruddy wine from silver flagons and echoed the laugh and song of revelry.

TIPPLE TOWERS: Searle and Herring's crumbling Castle Brewery pictured in 1893 (Gwent Archive)

Precisely at what time the poor ruin was seized upon by the base fortunes which now degrade it is not recorded. The humiliation of a modern red-tiled roof covers the central tower of the ruin where it looks upon the river, the only part of the walls now standing. The tower next the bridge has its venerable head covered with a kindly peruke of ivy, as if nature took pity on the poor old castle in its hour of degradation, and tried thus as it might to atone for man's inhumanity to storied stone.

Mr Sikes's cry of anguish from the heart was heard. In 1898 Newport rivals Lloyd and Yorath bought Searle and Herring and production ceased at the Castle Brewery, though it continued to be used as a bottle store until 1905. In 1891 the unused South Tower had already been acquired by Newport Corporation. After the closure of the brewery in 1899, the rest of the castle was purchased by Lord Tredegar. In 1930 the care of the ruins passed to the Ministry of Works and the remains of the brewery were cleared away.

Newport Castle had been cut free from the malt sacks and hop bines, but it never regained much grandeur, being wedged between the railway line built in 1850 and a road

widening scheme. It looks today as if it could do with a drop of the brewery's old vatted ale to cheer up the grim remains.

Exactly why Searle and Herring sold out is unclear. Chairman Arthur Evans had told the ninth AGM in May 1897 that the company had enjoyed a reasonably successful year with profits of £3,808 and a dividend of 7½ per cent. The decision to sell is more likely connected with Richard Searle's growing brewing interests elsewhere, some of which were not as successful.

In the 1890s the Searle family had taken short-lived control of the Monmouth Brewery. Richard Searle also had connections much further afield. He was a director of the newly-formed Newmarket Breweries in Suffolk in 1896, which went into liquidation the following year. Albert Herring was involved with the troubled Carmarthen United Breweries.

Lloyd and Yorath, which had taken over the Castle Brewery, was a new creation, having been formed in 1895 through the amalgamation of two Newport firms, wine merchants J.L. Lloyd and brewers William Yorath & Sons of the Cambrian Brewery in Baneswell Road. The joint venture was valued at £190,500. With the purchase of Searle and Herring, it became one of the largest drinks firms in the town. Its main rival was a noted brewing family from the Midlands, who when they came to Newport not only brought brewing expertise but also a strange new game with an oval ball – rugby.

Thomas Phillips had extensive brewing experience. With his brothers he ran the Phoenix Brewery in Northampton (later the Northampton Brewery Company). They had also established a brewery in Burton-on-Trent to brew pale ales, but in 1873 this was sold to Truman's of London. With his share of the proceeds Thomas Phillips decided to head for the land of opportunity – South Wales – where the population and industry were growing rapidly. In 1874 he bought the Dock Road Brewery with thirteen pubs from a former Mayor of Newport, Thomas Floyde Lewis.

When he arrived in Newport he brought with him ten children, and they brought with them, among the many chests and assorted baggage, a rugby football. This had been bought in the town of Rugby for 13s. 6d. and was claimed to be the first ever kicked in anger in the Newport area.

Two of his sporting sons, William and Clifford, soon set up a club after a meeting at the Dock Road Brewery in September 1874. They had planned to play soccer but could not find enough fixtures, so in April 1875 they played a match against one of the Cardiff rugby clubs. That fixture is regarded as the foundation of Newport's rugby team. William became captain of the famous club.

Away from the sports grounds, Phillips & Sons also prospered, branching out into wines and spirits. The leading son at the Dock Road Brewery was Edward. But with so many children to satisfy, Thomas Phillips briefly set up a second, smaller brewery in Station Street in 1883 with another son Frederick to supply the private home trade. After a year this distinct business was absorbed into the main company.

In 1886 a new 25-quarter brewery was built at Dock Road and on the father's retirement in 1892 a limited company was formed with a share capital of £100,000. In 1898 a substantial maltings was built at Penner Wharf. In 1902 new stables for thirty-six horses were erected. The brewery also used the Monmouthshire canal for deliveries, sending barges of barrels up through the fourteen locks.

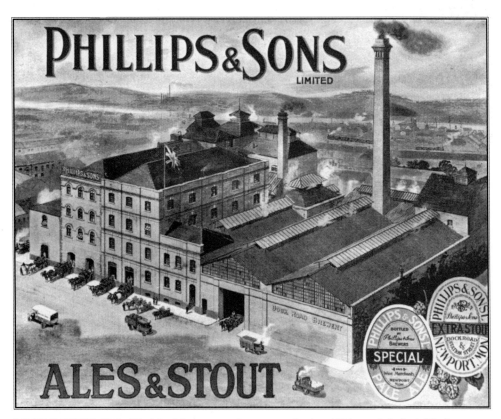

LARGER THAN LIFE: Phillips' Dock Road Brewery painted in exaggerated scale in 1924 to mark the firm's fiftieth anniversary. Bottled Special was one of its most popular brews (Courage Archive)

The company and the town prospered together. The brewery's managing director Frederick Phillips became Mayor of Newport in 1906, a post his brother Clifford had held in 1903. The capital of the company was regularly increased to £250,000 by 1918. This allowed the brewery to buy pubs including the extensive Pontnewynydd Brewery near Pontypool in 1912. By 1924, on the firm's fiftieth anniversary, there were 125 houses, some as distant as Haverfordwest and Hereford. The company claimed not to produce any specialities 'but its half-pint PSA is well known throughout Monmouthshire'.

Despite the progress the 1920s were difficult years for many breweries, and in 1925 Phillips seriously considered combining with local rivals Lloyd and Yorath, who also had over 100 houses. Only a last-minute 'hitch' in January 1926 prevented the amalgamation. Instead, just after the Second World War, both breweries fell separately to English companies eager to gain a larger share of the South Wales market.

The first to lose its independence was Phillips & Sons in 1949, when it was bought by Simonds of Reading in a share exchange deal. The large southern England brewer was pleased with its acquisition as, besides gaining 123 pubs, it filled a gap in its trading empire. It already had a depot in South Wales since taking over Stiles Brewery of Bridgend before the war.

Brewing continued at the Dock Road Brewery for nearly twenty years, producing at first Simonds AK and other draught beers (Phillips beers disappeared) and introducing a new bottling plant. However, it was hardly a promising start, as an internal report revealed after the first year's trading:

A wild yeast infection in the brewery, which started at the end of June and was not finally eliminated until October, had a disastrous effect on the trade and, coinciding as it did with an intensive campaign by Mitchells & Butlers and Ansells, it resulted in our losing a very large proportion of our club trade. In some clubs, where we had been established for years, we lost all trade.

Figures showed that business had dropped by 20 per cent in a year, down from 40,862 barrels in 1948 to 32,567 barrels in 1949. The strength of the Birmingham-brewed beers was also a problem.

To compete with this serious situation, a new and stronger beer was on sale as from October 17 with the express object of regaining club trade. Progress has been slow but there is no doubt that we are gaining ground, but the quality and gravity of the Birmingham beers make the task more difficult.

Simonds was taken over by Courage in 1960. A year later Courage added the Bristol brewery of George's to its empire. Once this plant was redeveloped and the Severn Bridge opened, allowing easier access to South Wales, the Newport brewery was doomed. It closed in 1968 and was soon demolished.

One of the feared Birmingham brewers, Ansells, snapped up Lloyd and Yorath (or Lloyds of Newport as it was known after 1946). The takeover was hardly surprising. The company had found it so difficult to compete with outside beers that Lloyd's had virtually turned its pubs into free houses. Chairman Walter Lloyd told the staff at the annual Christmas dinner in 1949 that the only way it could meet the stiff competition was by making many other products available – though he hoped that Lloyd's beers would still be the best sellers.

When Ansells made its offer in 1951 the dispirited directors immediately recommended acceptance. 'While the directors have a natural reluctance to see the company lose its local character,' they said, 'the balance of advantage is very definitely now with the big units.' After Ansells became part of Allied Breweries in 1961 the brewery closed, a depot being opened at Marshfield between Newport and Cardiff. The town-centre brewery was sold for a shopping development.

Not all the outside brewers had everything their own way. Birmingham giant Mitchells & Butlers had a surprise when it took over Thatcher's Bristol Brewery in Newport in 1949. The aggressive Midlands company thought it had just acquired twenty pubs and a useful base from which to push further into South Wales. Instead, it had stumbled on one of the most eccentric sidelines in British brewing history.

LOSING BREW: despite winning gold medals in competitions, Lloyd's beers found it difficult to compete with heavily-promoted beers from outside like Ansells

M&B bought the brewery in October, and in December district manager Edward Bennett was puzzled to see queues of expectant people gathering in Alma Street outside the brewery. Each of the arrivals seemed to be clutching cloth-covered basins neatly tied at the top. He sent someone outside to find out what was going on – and discovered that M&B had stepped into the heat of the Christmas kitchen.

A regular practice had grown up over forty years that the people of Newport could come to Thatcher's Brewery to steam their Christmas puddings. The company had a large steam chamber for cleaning casks that local housewives found was ideal for cooking their festive treat. The operation was no small beer. Christmas dinners all over town depended on it, with some 25,000 puddings steamed every December. Each one had to be left, labelled and collected a day or two later. During the Second World War the brewery safeguarded itself against the wrath of local cooks by pinning a notice over the chamber: 'We are not responsible for any war damage to Christmas puddings.' One local later recalled:

> Few residents would think of buying ready-made Christmas puddings, and few would consider cooking them at home, for it was far easier to take them to Thatcher's. And what was more important they tasted much better having been cooked in the environment of beer brewing! In any case it only cost 6d. per pudding.

Edward Bennett was dumbfounded:

I am amazed by the number. I don't know whether these good people think we put rum in their puddings. We have had to employ students to carry them into the place. The people have brought them here in anything from a perambulator to a Rolls Royce.

M&B retained the 'oven' solely for this practice for eleven years before the premises were closed in 1960. At a farewell party Mr Bennett commented: 'It is a service that we offered to the public for many years and we are sorry that the service had to end. I am sure it will be missed.'

He could have been talking about the brewing industry in Newport. By then all local beers had vanished.

HEART OF COAL

One other company might have joined Hancock's and Rhymney in the brewing league of the giants in South Wales. In fact in the mid-nineteenth century this venture was the largest brewery in the region, led in succession by three of the most formidable figures in the country's industrial development.

More than any other Welsh brewery, it was founded firmly on the black gold beneath the ground. This combination of coal and ale should have spelt success. Yet the Vale of Neath Brewery at Cadoxton, near Neath, had a chequered and fire-scarred history that set it apart. Its start was hardly auspicious. It was born in spectacular failure.

Nearby Swansea had been the cradle of the industrial revolution in South Wales, largely because the South Wales coalfield reached the sea at that point, making the coal more accessible and easily transported by boat. This ready source of fuel attracted copper smelters to the area. Near Neath appeared the Crown, Royal and Red Jacket copperworks. By the mid-1840s there were also nine ironworks. Coal mines were dug at a feverish rate to meet a growing export trade as well as local demand. Workers poured into the area, hungry for work – and after work, thirsty for beer.

A number of breweries sprang up to meet this need in Swansea. Buckley's was already brewing in Llanelli. But the most ambitious of all were Messrs Stancombe, Buckland and Rusher of the Maesteg Iron Company, who in 1836 instructed Swansea architect William Richards to build them a brewery at Cadoxton. A widespread trade was planned. The partners advertised for agents in Swansea, Cardiff and Merthyr. By 1838 they had begun producing ale and porter, and a year later were already claiming to be the largest brewing business in South Wales. Sales had reached 400 barrels a week. Someone must have become jealous because in 1839 they had to deny repeated rumours that they were adulterating their beer.

By the end of the year, in order to expand further, they created a joint-stock company with a capital of £100,000 grandly known as the Vale of Neath and South Wales Brewery Company. The trading area now covered the whole of South Wales into Monmouthshire. The public was invited to apply for shares at brewery stores stretching from Swansea to Newport. Two new partners, William Brunton and Joseph Little, were brought in. Plans were in hand to increase output to 1,000 barrels a week.

At the close of the company's first year in 1840 a dividend of 8 per cent was paid. Sales were said to have been good in 'higher quality' beers. The future appeared promising, but there were tell-tale signs of strain. In 1841 a new director, George Walters, a banker from Bath, was appointed. This was not to strengthen the board – but because he was owed £1,500 by the brewery manager, W.H. Buckland. A depression in the iron industry began to hit the business. Its better quality beers were especially affected.

Then in 1843 their crumbling empire went up in flames when a major blaze destroyed much of the brewery. The fire services were useless. The Neath engine could do nothing as it had defective hoses. The Swansea engine, delayed by lack of horses, arrived after the blaze had burnt out. Only rain prevented total destruction, but the brewery tower was gutted and the damage was estimated at £7,000. This was covered by insurance, but there was a damaging break in business that spelt ruin for a company burdened by fixed-interest payments to some shareholders. The brewery also owed £46,000 to the Maesteg Iron Company, with which it was closely linked.

After some dubious financial manoeuvres – which ended up being condemned in court – the company was dissolved in 1847. It had been 'a notoriously rotten concern', wrote a correspondent to the *Cambrian* newspaper, badly mismanaged by partners who knew little of the brewing business and who raised money by reckless and risky means. The ironworks inevitably followed, swiftly sliding into bankruptcy.

In October 1847 the brewery was auctioned at Bristol. Built massively of stone on a large site with a modern plant, it looked a good buy. But its troubled history and the depression of the period told against it. There were no bidders. A year later the plant – which had cost almost £15,000 – was sold separately for £2,700. It was not until the fourth attempt in 1850 that the buildings, including a three-floor maltings, were finally disposed of. Built at a cost of over £20,000, they were picked up for a bargain £1,520 by Evan Evans (who had earlier offered £6,000 for the plant and premises, but had been turned down). Control had passed to a shrewd man of business.

Evan Evans had been born at Ynysmaerdy near Briton Ferry in 1794. He trained as a confectioner, but soon showed a flair for wider business. By 1830 he had become licensee of the Grant Arms in Neath High Street, specializing in selling wines and spirits. In 1833 he gained a prized appointment, the local agency for Guinness. He advertised in the *Cambrian* in 1834 that not only did he have an 'unlimited supply of that inimitable beverage Dublin Porter' but also 'home-brewed Welsh ale, mild, clear and strong, in bottles and casks'. In addition he sold malt and hops.

His quick nose for business was sharply illustrated when the Neath races were revived in 1835. He put up a covered stand for spectators and provided a 'cold collation of prime roast and boiled beef, fowls, ham, sandwiches, jellies, spirits, bottled ale, porter, cider, soda etc.' His entrepreneurial spirit was demonstrated again and again.

With partners in 1846 and 1847 he bought two ships, the 99-ton steamer *Neath Abbey* and the 47-ton schooner *Liverpool Packet*. These were not only used for Bristol freight but also to run popular pleasure trips to Devon. One sailing in 1847 was so packed with 500 passengers at Neath, there was no room to take anyone else on board at Briton Ferry. When Brunel steamed into Neath on the first train in 1850, Evan Evans led the celebrations including the organization of an artillery salute to the new railway. His beer, of course, flowed freely.

By the early 1840s his Neath Brewery in the High Street was the main local rival to the much grander Vale of Neath Brewery at Cadoxton. He employed travellers to seek extra trade, one of whom, John Marandez, was found guilty of embezzling money in 1843. His barrels rolled further and further. In 1849 one of his draymen was killed in an accident after delivering beer in Ystradgynlais. Thus when he bought the Vale of Neath Brewery buildings in 1850, he was already experienced in the brewing business.

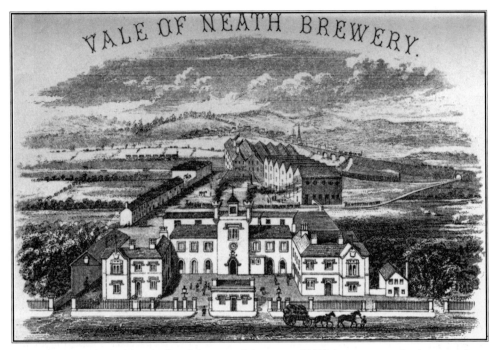

GRAND EDIFICE: a sketch of the Vale of Neath Brewery at Cadoxton around 1860 showing its unusual architecture and extensive site which reached back to the railway line

The premises were just what he needed. His High Street Brewery was cramped; here was space to expand, even if the buildings did need refitting. Unlike the previous owners, he knew what he was doing. Slater's Directory of 1858 was so impressed with the result, it unusually mentioned the company at length in its introduction to the town:

> On the entrance to Cadoxton from Neath is the Vale of Neath Brewery – an establishment worthy of notice – justly considered one of the most compact in the kingdom. A part of which was originally erected (1840) by a joint stock company at a cost of £60,000, whose fate like many other similar co-partnerships, was a ruinous failure; afterwards purchased by the present proprietor, Evan Evans Esq. who by skill and perseverance has refitted it throughout in a judicious manner, replete with every requirement, besides making considerable additions, by which it now covers upwards of four acres; an estimate may be formed from the weekly sale of 1,000 barrels.

If that figure is no exaggeration, it almost certainly made the Vale of Neath Brewery the largest in Wales at the time. Buckley's of Llanelli, for example, was not then producing even half that amount. And the firm did not stop expanding there. The 1861 census described Evan Evans as 'brewer employing 80 men'. In 1866 the brewery cashier told the *Cambrian* newspaper that the company employed 200 workers; in his eleven years service he had watched the brewery grow daily until it had reached its present 'enormous extent'.

Evan Evans was now much more than a brewer. He had been a town councillor since 1846

SHIPPING PORTS, SWANSEA AND NEATH.

EVANS & BEVAN,

NEATH,

SOUTH WALES,

Colliery Proprietors,

AND

EXPORTERS OF WELSH STEAM COAL

FOR LOCOMOTIVE, STATIONARY, AND MARINE ENGINES.

These Coals have for many years been in extensive use in England, Ireland, France (especially), Spain, and over the Continent.

The following are the descriptions of Coals Shipped:—

That superior and well-known Smokeless Steam Coal, extensively consumed by the English Government, and known on the Government List as the **GELLIA CADOXTON STEAM COAL**.

The **GELLIA CADOXTON STEAM COAL** has proved itself a first-class Fuel for working Steam Engines where Smoke is objectionable. It is highly esteemed on account of its great power of generating steam, and large quantities of it are used annually in the various Dockyards. It is much appreciated in FRANCE, and has received the marked approval of the Imperial Government. It is peculiarly adapted for use in Marine, Stationary, and Locomotive Engines, from the fact that it is entirely free from pyrites and other substances, which render some of the South Wales Coals objectionable.

EVANS & BEVAN'S THROUGH AND THROUGH COALS, about two-thirds large, used for working Steam Engines where smoke is objectionable.

EVANS & BEVAN'S SMALL COAL, so much appreciated and consumed at the various patent Fuel Works.

EVANS & BEVAN'S FAR-FAMED BITUMINOUS COAL is well adapted for General House purposes, Smiths' Work, &c.

EVANS & BEVAN'S CULM is extensively used for Burning Lime, &c.

Messrs. EVANS & BEVAN, from the arrangements which they have entered into, are enabled to guarantee quick despatch to vessels of any size, and can ship at the rate of EIGHT HUNDRED TONS per diem.

COAL CONNECTION: an advert for Evans & Bevan's extensive mining interests in 1868

and was elected Mayor of Neath in 1856 and 1862, despite the fears of some about having a brewer as Chairman of the Watch Committee. From 1864 he was an Alderman. When the Prince of Wales married in 1863, the brewery dray-horses felt the full weight of the pageant, some having to carry knights in armour while others pulled the artillery for another gun salute. He established an efficient fire service in the town (a wise precaution after the earlier fate of the Vale of Neath Brewery) and in 1860 formed the 17th Glamorgan Volunteer Rifle Corps.

He loved a good show, but was not afraid to get involved in more vital matters. When cholera struck Neath in 1866, he bought lime and organized brewery workers on a dawn raid to quicklime all the water courses and drains in Cadoxton. The life-saving action took place between four and six o'clock in the morning so as not to alarm the villagers.

For such a forceful man he was a diplomat when necessary. With the temperance movement growing in strength, he kept the right side of the Church through charitable gifts. He even erected a chapel at Bryncaws. When the Welsh Independent Association visited Zoar Chapel in 1865, he provided free dinner and tea at the Vale of Neath Arms for the

100 ministers who attended. After building the imposing Jersey Marine Hotel on the coast in 1864, he loaned its extensive facilities free of charge for Sunday school treats.

He was a philanthropist in other fields. Every Christmas the poor flocked to his office for fourpenny pieces. In 1850, when the failure of the Briton Ferry ironworks caused hardship, he set up a soup kitchen to feed the unemployed.

Evan Evans could afford to be generous. He was becoming an industrialist on a grand scale. In 1864 he bought land to the north of Neath for coal mining. This venture was in partnership with his son-in-law David Bevan, a trained surveyor and colliery engineer, who had married his daughter Mary in 1846. This combination of Evan Evans's capital and business acumen with David Bevan's expertise proved fruitful. By 1867 Evans and Bevan owned four mines – Gellia and Rhydding near Cadoxton and Torcefn and Crynant in the Dulais Valley. As a shipping merchant, Evan Evans was in an ideal position. Soon 800 tons of coal a day were being exported, particularly to France.

In the late 1860s he also expanded his brewing interests by buying the Pontardawe Brewery from Robert Evans, some eight miles from Neath. Renamed the Swansea Vale Brewery, this was managed for many years by John Evans.

When Evan Evans died in 1871 he was known as 'the grand old man' of Neath. His life and business touched all aspects of the town and everyone turned out for his funeral. Shops and businesses shut, and church bells tolled, as a long procession wound through the crowded streets led by his rifle volunteers. For a humble confectioner he had made a huge impact.

David Bevan, who inherited Evans's vast business empire, was a man of no less stature, though probably more comfortable with the mining trade than the brewing business, despite having once been burnt in a pit explosion at Cwmavon. He lost little time in opening another two collieries, bringing the total to six. On 11 March 1872 the first coal trucks steamed away from Brynteg down the Neath and Brecon Railway, while on the same day Isabella, the eldest of his seven daughters, cut the first sod of the Seven Sisters Colliery.

As one of the leading coal owners and brewers in South Wales, he followed a path remarkably similar to his father-in-law. A town councillor in 1869, he became an Alderman in 1871 and Mayor of Neath in 1872 and 1873. He is said to have left his mark on local government history by being the first mayor to wear civic regalia that he bought himself.

After his death in 1888, his only son Evan Evans Bevan took over, not only giving his name to the brewery but also devoting much more time to the beer business. When a commercial journalist called in 1893 he was shown personally round the High Street offices by the proprietor and then driven out by him to the brewery in Cadoxton.

Evan Evans Bevan must have been troubled by the temperance movement, for he bent the reporter's ear with a remarkable argument which was repeated in the writer's gushing profile of the Vale of Neath Brewery:

If all the beer offered for public sale were of similar quality to that supplied from this brewery, good and wholesome, there would be less heard of the grievous drunkenness which so disgraces our land. The consumption of adulterated mixtures dignified by the name of beer must be held responsible for a great deal of this blot upon our national reputation.

When the reporter interviewed Mr Bevan, he was already Mayor of Neath, a post he was to hold four times, including three years in succession from 1902. He also served as a councillor from 1889 to 1905, but did not let his civic duties get in the way of his industrial concerns.

He bought and built pubs at such a rapid rate that by 1919 the company owned 148 houses, most in the industrial valleys around Neath. A valuation that year estimated that the company was worth almost half-a-million pounds. The Vale of Neath Brewery by now covered 11 acres with its own private railway siding. There were six motor lorries and eight horse-drawn drays. It was a healthy business. The bulk of its trade was in large barrels, each holding 36 gallons of beer. Its cellar stock in 1919 was 436 hogsheads (54-gallon casks), 1,906 barrels and 792 kils (18 gallons). There were only 81 smaller firkins (nine gallons) and 50 pins (four-and-a-half gallons). And brewing was only part of his business. On Evan Evans Bevan's death in 1929 he was the leading colliery owner in the country with seven mines. He was also said to be the largest individual employer in Britain. The Onllwyn Colliery, for instance, employed 1,200 miners. He also owned large estates in Breconshire.

His only son David Evans Bevan took over this extensive empire and in 1935 turned the

brewing side into a limited company, Evan Evans Bevan Ltd, by now with 207 pubs. Two years later the much smaller Swansea Vale Brewery at Pontardawe was closed, with production concentrated at Cadoxton.

The same handful of directors also ran the mining company, Evans and Bevan. This close combination of beer and coal led to a number of dubious practices, like allowances being paid to the men in beer chits rather than cash. The afternoon shift was always allowed up an hour before 'stop-tap'. A later director of the South Wales coalfield even claimed: 'It was alleged too that they deliberately restricted the ventilating current to ensure that the men would be thirsty at the end of the shift, but I have no proof of this.'

In 1947 this link was broken when the coal mines were nationalized. This allowed wealthy David Evans Bevan to concentrate on the brewing business. More pubs were bought and several hotels built. Output at the brewery was trebled through the installation of modern equipment including new coppers, fermenting vessels and bottling plant.

In 1952 the company took over David Williams' brewery at Builth Wells. The brewery, founded in 1866 by John Prosser, was closed, but a depot retained in this distant rural area of central Wales. Andrew Buchan's Breweries of Rhymney had considered buying David Williams the year before, but had been put off by the poor state of some of the thirty-three pubs.

In 1954 David Evans Bevan's son Martyn became a director, but gradually the family seemed to lose interest in the business. In 1958 the father was made a baronet, becoming Sir David; in 1967 he decided to sell out to Whitbread. He said it was due to taxation problems with the privately-owned company. It was an odd, off-the-cuff deal.

Whitbread had already taken over the larger Rhymney Breweries the year before, but although the Vale of Neath pubs sold Whitbread bottled brews like Mackeson and Forest Brown, the London giants had no share stake in the Welsh company or close connections. That all changed on a train journey from Swansea. Col. Bill Whitbread met Sir David and agreed to buy the brewery, with its 240 pubs. The move so surprised Col. Bill's fellow directors that £4 million had to be rapidly borrowed from Barclays Bank to fund the purchase.

A Vale of Neath representative commented when the news broke on 17 June: 'Sir David really made a point of asking them (Whitbreads) if they would give a definite assurance of a continuation of employment for all his employees and Whitbreads agreed.' There were 200 workers. A Whitbread spokesman said brewing would continue at Neath.

But there was a doomed feeling in the air. The brewery that had been born with a blaze was about to end with another fire. Before the deal had even been completed there was a major inferno at the Cadoxton brewery on the night of 22 August. Firemen rushed from Port Talbot, Pontardawe, Briton Ferry, Pontypridd, Neath and Bridgend as flames leaped 80 ft from a five-storey building. Ninety minutes after the blaze was first discovered the roof and a number of floors had collapsed. 'At one time firemen were battling desperately to prevent the flames reaching the spirit stores where hundreds of gallons of spirits were stacked', reported the *South Wales Echo*.

The fire had not destroyed the brewhouse but the bottling plant and keg store. Draught beer was not affected and Whitbread still took control on 1 September 1967, but it was the beginning of the end for the Vale of Neath site. Inevitably bottling was concentrated at Cardiff, and in 1969 Evan Evans Bevan and Rhymney Breweries were merged to form Whitbread Wales.

On 31 May 1972 the Vale of Neath Brewery finally shut, with the loss of 140 jobs. The company that had chosen a heart for its symbol had suffered its final heart attack.

WASHING AWAY THE DUST: a Rhondda miner, tin jack under his arm, enjoys a well-earned glass of beer at the New Inn, Ton Pentre, in 1906

WONDERFUL WEBBS

In the eastern valleys, alongside Rhymney Breweries, the main pit brewer was Webbs of Aberbeeg. Its closeness to the coal-face is shown by its early logo depicting picks and shovels around a barrel perched on a pile of coal.

PICKAXE PINT: an early Webb's logo shows the close connection between brewing and mining

The family were major land owners in the area and ran a number of related companies from malting to wine and spirits merchants. Founder William Webb of Llanhilleth House was also a colliery proprietor. Although his ventures date back to 1838, brewing does not appear to have started on a large scale until the 1860s at Aberbeeg, which was strategically sited at the junction of a number of valleys alongside the railway.

The firm traded as Webb Brothers in the second half of the nineteenth century before being registered as Webbs (Aberbeeg) Ltd in 1905. In the same year the brewhouse was considerably enlarged giving a total capacity of 1,460 barrels. The premises were expanded again in the early 1950s after taking over two Rhondda breweries, David John of Pentre in 1946 and the Fernvale Brewery, Pontygwaith, in 1949.

In this same period the company expanded into cider production, taking over Ridlers of Hereford (1944) and William Evans of Hereford and Devon (1949). Sales of cider, which was cheaper than beer, were always substantial in the mining valleys, particularly during depressed periods.

In 1960 the company was taken over by United Breweries of Great Britain, a company formed by the Canadian Teddy Taylor to take over breweries to market his Carling Black Label lager in Great Britain. United joined with London brewers Charrington in 1962 and in 1967 this new group linked up with Bass to form Bass Charrington.

When Bass took over Hancock's in 1968, Webbs was merged with the Cardiff brewers to form Welsh Brewers, the Aberbeeg brewery stopping production in 1969.

PINT AND THE PULPIT

Everywhere in Wales the chapel and the pub glare at each other from opposite street corners, or even sit uneasily side by side in the same terrace, the chapel gazing sternly past its more rowdy neighbour like a refined gentleman trying to ignore the tramp on the same park bench. But in one brewery the pint and the pulpit sang together in harmony.

Buckley's Brewery of Llanelli owes its early years to a remarkable entrepreneur, Henry Child. Born in Freystrop, near Haverfordwest, he came to Llanelli in 1760, aged eighteen, to act as estate agent for local landowner, Sir Thomas Stepney. But soon he was striking out on his own.

Encouraged by the town's industrial expansion and growing population, he leased the Talbot's Head inn in 1769. Shortly afterwards he bought an old malt-house and then another pub, the Falcon, at the bottom of Thomas Street (today still standing as offices next to the brewery). From the Falcon he ran his estate business, receiving rents there, developed a busy market round the walls of the neighbouring parish church and held auctions at the pub. From the profits he leased the Carmarthen Arms in Wind Street in 1791 and then built the White Lion nearby.

All these pubs, brewing their own beer, provided a ready market for his malt, and in 1799 he took the next logical step, obtaining a 55-year lease on the central site on which the brewery now stands. Determined to control the whole business from the ground to the pot, he also leased farms to grow his own barley. That other essential of life, bread, did not escape his attention, and he took over the local flour market, leasing the Llanelli Mill and buying the Felinfoel Mill. This was followed by taking a wharf at Llanelli Dock to export his surplus grain and flour.

The country lad from Pembrokeshire had arrived. In 1795 he went into partnership with Lord Ashburnham to work Pembrey Colliery, and in 1810 he was named trustee in the Act of Parliament for enclosing lands in Llanelli in order to improve the town. But what made him stand out from other entrepreneurs of the age, apart from his resounding success, was that at the same time as he obtained his first pub – he became a Methodist.

John Wesley visited Llanelli in 1769, and Henry Child may have been converted by the great preacher himself. He went on to become a leader of the local Methodist Society, building the town's first chapel, Wind Street Chapel, in his own garden in 1792. John Wesley often stayed in his home, as did many other itinerant preachers. Among them was the Reverend James Buckley – whose introduction to the area was nearly fatal.

James Buckley was born in 1770 in Oldham, Lancashire. His father, Mark, was a weaver who later turned his house on Sholver Moor into a pub called The Trough (now the Wagon and Horses). By the age of fifteen, James was already holding prayer meetings, a bold move for one so young. Once, in a neighbouring village, he was almost thrown into a lime-kiln by a mob. By 1791 he had become a Methodist minister, and two years later was appointed to the Glamorganshire circuit, which stretched from Chepstow on the English border to Llanelli.

Travel was then a hazardous experience, as James Buckley discovered when he attempted to reach Llanelli in 1794. The *Wesleyan Methodist Magazine* of April 1842 records:

> Near the end of his journey he had to cross a small arm of the sea which was fordable at low water. A person whom he saw there told him he might cross with safety and directed him to keep a certain object steadily in view. Mr Buckley had not proceeded many yards, when the water became so deep, and the current so strong, that he was carried downwards about a quarter-of-a-mile, before he reached the other side, and then he only found a soft mud, insufficient to bear his weight.

His life might have ended there. 'Happily, a person who was perfectly acquainted with the area saw him, called upon him to stand still, and then hastened to him and guided him out of his difficult and most perilous position.' This estuary escape is commemorated in the name of one of Buckley's nearby pubs, the Reverend James in Loughor. 'Completely wet and covered with sand and mud, he got over the remaining four miles of his journey as quickly as possible, and rejoiced to find himself under the hospitable roof of Mr Child.'

It was there that he met Mr Child's daughter, Maria, whom he married four years later in 1798. But the couple did not settle in the Principality, instead moving from one Methodist mission to another in England. They did not return to Wales until 1823, a move prompted by Mr Child's growing ill health. When Henry Child died a year later, James Buckley found himself with two concerns to run – saving souls and satisfying thirsts.

He tackled the problem by concentrating on his ministry, based in Swansea, while leaving his sons to look after Henry Child's wide business interests. It was not a solution without problems. In one letter to his eldest son, Henry Child Buckley, he threatened to sell up if Henry did not stop 'guzzling and drinking to excess repeatedly'.

There seems to have been no escape for Reverend James from the demon drink. When he was briefly appointed to Carmarthen from 1827–9, he found his chapel there had been built above a brewery cellar. This odd combination inspired the following verse:

> Spirits above and spirits below,
> Spirits of bliss and spirits of woe.
> The Spirit above is the Spirit Divine,
> The spirit below is the spirit of wine.

James Buckley died while attending the Methodist Centenary Conference in Liverpool in 1839. His body was brought back to Llanelli and buried in the parish church – across the road from the ever-expanding brewery. There was no freedom, even in death, from the pint and pulpit connection. Over 150 years later, Buckley's Brewery would produce a premium draught ale called Reverend James to celebrate the remarkable time when the glass and the gospel prospered together.

Reverend James's second son, also called James Buckley, took over the brewery. Born in 1802 he was the epitome of the Victorian businessman, ruling with a rod of iron for almost

BEER AND BIBLE: the close connection between the church and Buckley's Brewery was celebrated in 1991 with the launch of a premium ale called The Reverend James

as long as Queen Victoria sat sternly on the throne. From his letters he comes across as a man who was hard on his staff, family and himself. Intolerant of advice that went counter to his own ideas, he was constantly seeking after profit, yet all his dealings were laced with high moral overtones.

He was a figure who was respected, even feared. When a son-in-law, who managed a small bank in Llanelli, hit difficulties owing to a run on the bank, James Buckley called out one of the brewery drays. Then flanked by his staff, he marched in a proud procession to the besieged building, where his draymen ostentatiously unloaded bars of gold and carried them into the vacant vaults. The crisis was ended.

James Buckley was so unyielding to one of his senior staff that the aggrieved employee left and set up a rival brewery. J.Innes in his book *Old Llanelly*, published in 1902, recalls: 'No Llanellyite ever thinks of this huge concern (Buckley's) without remembering the energy and ability Mr William Bythway applied for so many years to its interests.' No Llanellyite might forget Mr Bythway's services, but James Buckley was not impressed.

In a long series of letters, William Bythway, the manager of the brewery, constantly complained about his salary and conditions of employment. Tiring of making no progress on this front, the enterprising Bythway switched tactics and proposed to one of Mr Buckley's daughters. His advances were met with a withering rejection and so, snubbed at every turn, he walked out and in 1875 set up his own brewery in the town.

James Buckley's long reign came to an end in 1883. He was succeeded by his two sons, James and William Joseph, who traded as Buckley Brothers. Under them the brewing giant of South-West Wales was to become a limited company. But before that happened, they were to have a distinguished visitor.

It is a mark of the high standing of Buckley's that Alfred Barnard chose the Llanelli concern as one of only four breweries in Wales covered in his monumental four-volume work, *Noted Breweries of Great Britain and Ireland*. The others were Brain's of Cardiff, Pont-y-Capel near Merthyr Tydfil and Soames of Wrexham.

On his visit in 1890 he commented on the 'lofty and most substantially constructed' buildings on the four-acre site. In the three spacious store cellars he discovered 'nearly 7,000 barrels of ale'. Another four-storey building 'recently purchased by the firm' was to 'warehouse 5,000 barrels'. Barnard was particularly impressed by the 'splendid stud of 18 horses, which has no equal in Wales'. The tour of the premises took so long that it was conducted over two days, with a visit to the extensive maltings, 'the largest malt works in Wales', on the second day.

When Buckley's Brewery Limited was formed in December 1894, the whole business was valued at £162,350. The company directly owned or leased 120 pubs and many others were tied by loans. The average barrelage for the three previous years was said in the prospectus to be over 23,000 barrels a year. By this time Buckley's had agencies selling its beer throughout

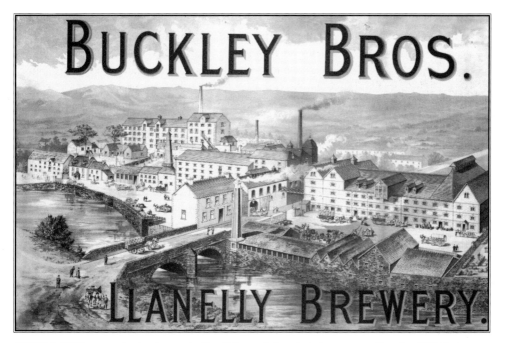

SMOKING AWAY: a poster showing the extensive Llanelli brewery before a limited company was formed in 1894 (Crown Buckley)

South-West Wales in Tenby, Newcastle Emlyn, Llandysul, Carmarthen, St Clears, Ammanford, Mumbles and as far east as Maesteg.

Soon Buckley's Brewery was to expand further by taking over two of its local rivals. But even as negotiations were under way, the fledgling limited company suffered a blow when the senior partner, James Buckley, died unexpectedly, aged fifty-eight, in September 1895. The company lost not only the leading family member but an influential figure in the region. James Buckley was the High Sheriff for the County of Carmarthen. He left a personal estate valued at over £65,000, including his home Brynycaeran Castle.

His younger brother, William Joseph Buckley, was left with a thorny problem on his hands. In the spring of 1895 William Bythway had indicated through his bank manager that he was prepared to sell his New Brewery. But, not surprisingly for a man with bitter memories of the Buckley family, he was proving difficult, refusing to provide firm details about the barrelage of the brewery or the size of its estate. Mr Bythway even suggested that Buckley's should say what it would give for a certain barrelage and a certain number of pubs in order to fix a price.

This was not acceptable to Buckley's, but the pressure was mounting with talk of a syndicate being formed to buy Bythway's and two other breweries in the county. Eventually W. J. Buckley offered £70,000 for 13,000 barrels a year. Mr Bythway then requested a private meeting at which he claimed, somewhat unconvincingly, that though he had offers 'from other quarters' he wanted to give Buckley's first refusal owing to its past attachment. 'A lot of other sentiment was introduced by Mr Bythway', it was noted by Buckley's solicitor.

A further meeting then followed to close the bargain. At this head to head in Mr Bythway's home in December 1895, Buckley's claimed that Mr Bythway said the brewery averaged 259 barrels a week and had 90 tied houses. On this basis W. J. Buckley upped his offer to £80,000, which was accepted. William Bythway was to take £10,000 worth of shares as part payment and to become a director of Buckley's.

'Then Mr Bythway's solicitor produced a schedule of the properties . . . and without parting with the same he turned over the sheets running his finger down the columns', noted Mr Buckley's solicitor. 'This production of the schedule lasted about one minute, perhaps less.'

When, days later, after signature, Buckley's finally got its hands on the figures, all was not well. Many of the pubs were on shorter leases than expected and some were 'of little or no value'. The barrelage figures were cleverly averaged over six years to boost the total. In 1891 the New Brewery had produced over 15,000 barrels, but the trade had 'considerably diminished during the past four years'. By 1895 Bythway's annual business was under 12,000 barrels. A stark assessment of its new assets by Buckley's valued the bricks and mortar of the brewery premises, plant, pubs, drays, casks etc. at £19,450.

Mr Bythway, it appeared, had after twenty years gained his revenge. Buckley's called for compensation and asked legal counsel if it could repudiate the contract, but eventually seems to have dropped the case, a final price of £74,000 being agreed. Mr Bythway never became a director.

Despite the tricky twists and turns, the takeover, completed in March 1896, with a further £60,000 share subscription, was still good if a little expensive for Buckley's. It had gained 85 leasehold pubs which, with other purchases, brought its estate to 216 houses. More importantly, concentrating the combined production of over 36,000 barrels at its brewery brought considerable savings. At the AGM in December 1897, chairman W. J. Buckley reported that the purchase had 'proved a valuable addition to the business'. There were more misgivings about a second takeover.

The ancient market town of Carmarthen, north-west of Llanelli, boasted two major breweries in the late nineteenth century – Norton Brothers of the Carmarthen Brewery, Springside, and David Evan Lewis and Son of the Merlin Brewery in John Street. In 1890 the two combined as Carmarthen United Breweries Limited under the chairmanship of Philip Septimus Dowson, a prominent figure in Welsh brewing and mining circles (including being a director in the Nundydroog gold mine). He was a former partner in Dowson Brothers' Phoenix Brewery of Cardiff, which had been taken over by William Hancock in 1888. He then joined the board of Hancock's and one of his fellow directors, Joseph Gaskell, was also involved in the formation of CUB.

London accountants Alexander, Daniel, Selfe & Co. gushingly predicted in the prospectus: 'We have much pleasure in stating that both breweries appear to have been well managed, that their business has increased, and that in our opinion a promising and profitable future is before them, especially seeing that the combination will promote alike efficiency and economy.' These words were to ring hollowly through the next few years, as United creaked from one crisis to the next.

The first year began well with a 10 per cent dividend for shareholders. It proved to be a false dawn. At the second AGM in 1892 the directors offered a drayload of excuses instead.

'Owing to many unforeseen circumstances, the saving anticipated has not been attained in last year's working', they declared. 'Unusual delay occurred in the completion of necessary alterations to the brewery plant. From this and other causes considerable returns of ales resulted. The costs of successfully appealing against the magistrates' refusal to renew several licences, a wet summer and dear hops also adversely affected last year's profits.'

One angry shareholder criticized the directors for 'making improvident purchases'. The number of pubs now stood at over sixty, with twenty-five having been bought during the year. The directors admitted that the bank overdraft 'was rather heavy'. In response the board ended the dual management of the old firms and took direct control themselves. It did little good.

At the third AGM in 1893, the chairman Mr P.S. Dowson was so nervous that he ordered the proceedings be conducted behind closed doors, with the press outside. But the bad news leaked out. A feud seemed to be developing between the old firms. Mr Marshall, brother-in-law of director Col. Reuben Norton, attacked a circular issued with the report as 'impertinent'. The chairman regretted its issue. Another Norton, Owen, jokingly offered to sell his shares. It was a comment that was to have considerable significance later.

By 1896 Mr Dowson had retreated to London, leaving his address as the Saville Club, Piccadilly. Thomas Jenkins chaired the AGM, but the news was the same – 'the directors very much regretted having to report such a bad result of the year's trading'. A new manager and a fresh brewer were appointed.

In 1897 the festering boil burst. David Lewis, one of the former owners of the Merlin Brewery, presided, and he was angry that brewing had been concentrated at the Carmarthen Brewery. The report in the *Brewers' Journal* caught the bitter mood.

The chairman said many things had happened which had thrown a cloud over their concern, and they would not get right until it was thoroughly adjusted. He mentioned among others the spot chosen for the brewery, in spite of his advocacy of its being at the Merlin. He could not state exactly the loss, but he put it down at about £800 a year, and now the directors saw he was right and would be glad to make the change if the necessary funds were available.

The directors who had seen the light no longer included a Norton. Col. Reuben Norton, along with the previous chairman Thomas Jenkins, had resigned. Among the new directors was Albert Herbert Herring of the Castle Brewery in Newport.

The new chairman said that he was 'heartily sick and tired of other schemes, as those which they'd had on hand for the past two years had had a very detrimental effect on their company, and after a great deal of trouble and expense had resulted in nothing'.

The Nortons stood accused of a graver offence, added David Lewis:

When the businesses of Norton Brothers and D.E. Lewis and Son were bought by the company, they never thought that any of the vendors – who had a large stake in the concern – would open up a business in ale, spirits and mineral waters in opposition to them. This had been done at Carmarthen, and it would be difficult to calculate the loss and bad effect this had had on the company's concern.

He was referring to George Talbot Norton (one of United's original directors) and Owen Norton who, with the help of their family, had established themselves as beer, wine and spirit merchants in Priory Street by 1895. Among premises inherited from Henry Norton, who had died in 1893, was also the former Vale of Towy Brewery in nearby St Peter's Street. This had been Norton's mineral water factory. Now it was turned back to its original purpose and by 1899 was brewing again as Owen Norton & Co.'s Vale of Towy Brewery in direct competition with CUB.

United was the last word to describe the troubled Carmarthen concern, and it was little wonder that Buckley's viewed CUB with deep suspicion. In fact it took the Llanelli firm six years to overcome their doubts.

As early as March 1894 Lionel Taylor was writing to Buckley's from a firm of Cardiff stockbrokers: 'As you may be aware, the above brewery (CUB) has been put in our hands for disposal, and . . . as you are the people most likely to benefit by acquiring this property, we shall be glad to know if you are inclined to entertain the idea.' The asking price was £62,000.

Four years later Taylor was still bombarding Buckley's with particulars. 'Your directors are doubtless fully aware of the working of this brewery for the past three or four years, and know the reason why it has not paid.' Three days later, he added, 'It is capable of doing very much better.'

By now Taylor was becoming desperate. On 25 March 1898 he pleaded: 'Make us any reasonable offer, we will put it forward. It may result in a sale.' On 31 March he boasted that he had won a 'big reduction' – down to £50,000.

Buckley's must have been tempted at this price into making an offer for part of the firm, for on 22 April Taylor warned that he could 'only deal with this matter on the basis of a bid for the whole property'.

However, CUB was not in a position to haggle. Trade by 1897 was already down to 8,000 barrels a year. Eventually, by the end of January 1900, a deal was struck. Significantly, given United's past problems, the CUB directors had to agree not to set up as brewers, maltsters, wine and spirit merchants or mineral water manufacturers within a 25-mile radius of Carmarthen. The chairman, David Lewis, joined Buckley's board.

CUB, through the Norton family's interest in coal mining, had a handful of pubs as far east as Merthyr Tydfil and a depot in Aberdare, but this distant trade was disposed of by Buckley's, who concentrated on consolidating its position as the brewing giant of the South-West. The brewery in John Street, Carmarthen, was used for many years to manufacture Buckley's mineral waters using water from 'the well-known Merlin Well'. The maltings also remained in operation.

The ultimate seal of approval was fixed to Buckley's business in 1903 when it was granted a Royal Warrant of Appointment as brewers to the Prince of Wales. By this year the company was making annual net profits of £18,750, and its reputation was soaring.

These were golden years for Buckley's in every sense. In 1904 its beers won the

Champion Prize Gold Medal at the Brewers' Exhibition in London, and followed up by taking the Gold Medal for Ales and Stout at the Paris Exhibition in 1911. In between these victories it thrice won silver medals. To crown it all, when the Prince of Wales ascended to the throne in 1910, Buckley's became brewers by appointment to King George V. No other Welsh brewery earned this distinction.

ROYAL CONNECTION: a Buckley's advert of 1928 showing its bottle label boasting 'By appointment to His Majesty the King'

The First World War pushed the company to the limits. Buckley's whole fleet of motorized drays was sent to assist the British forces in France. Altogether 64 employees were called up, nearly half of whom were wounded or struck down by fever. Five were killed including one of the directors, Captain Hume Buckley Roderick.

After the war the firm expanded again, acquiring pubs from the Cardiganshire coast to the Swansea and Neath Valleys. In 1924 the capital of the company was increased to £240,000.

DEPRESSION BEATER: Buckley's try to beat the growing economic crisis in 1929 by selling its Golden Ale at 4d. a bottle

Much of the success of this period can be attributed to the shrewd management of William Thomas, who worked in the brewing industry in Llanelli for a remarkable sixty-four years. He first started under James Buckley in 1864, later joined Bythway's Brewery, and became manager of Buckley's in 1896 when the two firms merged – a position he held until forced to retire owing to ill health in 1928. Buckley's chairman Martin Richards commented that Thomas's 'unswerving devotion and transparent honesty played a foremost part in bringing into fruition the success and prosperity which the company had for years enjoyed'.

The Buckley family also played their part, particularly during the lean years of the early thirties. William Howell Buckley told the company's bowling club dinner in 1936 that there had been 'many tempting offers to buy the brewery', but his father had always told him 'whatever

Rolling Out The Barrels

*M*ost breweries make an impression on their home town, particularly if they are in the centre. But Buckley's had the power for years to bring Llanelli's traffic to a complete halt.

Gilbert Road, once a major thoroughfare, ran right through the middle of the brewery. Buckley's partly overcame this splitting headache by a bridge. As Alfred Barnard remarked on his visit to the brewery in 1890: 'The brewhouse, which is a lofty and spacious building, is connected with the fermenting department by an iron bridge erected over the roadway.' A century later it is still there.

But the bridge was not able to carry everything that needed to be moved between the two halves. Beneath the brewhouse was the cask washer, and after the casks had been cleaned they needed to be transported across to the fermenting block to be filled. Buckley's tackled this problem by literally (as pictured) rolling out the barrels — a couple of hundred at a time — across the road.

Harry Davies in his book Looking Around Llanelli *recalled the bizarre sight:*

It was a simple yet skilful operation, the barrels being manually rolled on wooden skids placed end-to-end across the road

from the door of the wash-house to that of the racking department, where they went down a chute at great speed. It was an eye-opening experience to see how Mr John Chester, posted in a basement on the further side of Gilbert Road, in one smooth movement caught, twisted and rolled away each barrel as it hurtled at him. This was a job demanding timing and concentration, and which was even more exacting when barrels were of oak and the brewery used massive 54-gallon hogsheads.

This exercise happened two or three times a day, lasting 15–20 minutes each time with occasional stops to let the traffic through. Motorists were not always amused, and a borough planning map of the 1950s went so far as to describe the central position of the brewery as 'an unfortunate site'. Drivers finally escaped the barrel barrier in 1970, when as part of a major £250,000 reconstruction of the brewery, cask washing and filling were placed under the same roof.

BUCKLEY'S BITTER

GROWING CONCERN: the Llanelli premises during the reign of the second James Buckley up to 1883 (Crown-Buckley)

you do, don't sell the brewery.' He added: 'There have been difficult times, but we have weathered the storm.'

As the drinking trend switched from draught to bottled beers between the wars, Buckley's leading brand came to be their bottled Special Welsh Ale – proudly bearing the royal coat of arms. It flew the flag not only in South Wales but around the world, as the *Llanelli Star* reported on 25 February 1928:

> Yesterday afternoon great interest was aroused at the sight of several brewery lorries proceeding to the station loaded with huge crates containing bottled beer which was labelled for India and South America. Enquiries by a *Star* representative revealed the fact that this was the first large consignment of bottled beer to be sent to these places from Llanelly. The beer was supplied by Messrs Buckley's Brewery.

The fame of Buckley's was spreading, distributing beer not only to India and South America (particularly Argentina), but also to South Africa and Australia. But within a few years, as the Depression took the gloss off most businesses, it was beaten to a major innovation by its local rivals, the Felinfoel Brewery.

KINGS OF THE CAN

elinfoel Brewery is similar to many other small family breweries once scattered across
the country – except in the surprising scale of its achievements. Felinfoel was the first
to can beer in Europe (and almost the first in the world). The brewery won Britain's
top prize for its cask beer. And, most surprising of all, it survived when almost all other
breweries of its size in Wales were taken over and closed down. But it was its pioneering
work in beer canning which was to have a lasting impact on the brewing industry, and its
early connections give a strong hint of its later enterprise.

Founder David John owned iron and tinplate works around Llanelli when, in the mid-
1830s, he bought the King's Head opposite his home in the village of Felinfoel. This was no
ordinary pub but an important coaching inn with its own blacksmith's shop to service the
horses. It also had a more worrying feature. Alongside the building, which stuck out into the
road, was a toll gate.

In the late 1830s this was like a red rag to an angry bull. The Rebecca rioters, the hard-
pressed tenant farmers who bitterly resented the road charges on their wagons and animals,
were rampaging through Carmarthenshire destroying toll gates. Feeling was running high
against the authorities. David John decided to abandon the king and renamed his tavern the
Union Inn.

Like most pubs of the time it brewed its own beer. This proved popular and soon his
Felinfoel ale was being sold to other houses. Then, as demand increased, he erected a larger
brewery in 1878 opposite the pub in the grounds of his house, Pantglas. The imposing stone
brewery was erected on his orchard, leaning right up against the road.

The premises became a focal point for the community, employing about fifty people. A
villager later recalled:

Nearly every family kept a pig in their garden. When the butcher was booked to kill a
pig in the back yard, large cans of hot water were carried from the brewery to scrape

and clean the pig. Some people living in the vicinity even carried hot water for their weekly washing. Ladders were borrowed, tools sharpened, any excuse to go into the brewery for a drink.

On brewing day, farmers from the surrounding area came for the 'sog', the spent grains from the mash tun to feed their animals. 'The yards and road outside were crowded with the farmers' carts, waiting their turn to be served.' The narrow road came to be known as Farmers Row.

Felinfoel Brewery was actually built astride the River Lliedi, which still runs through the premises today, and in whose waters hundreds were baptized by the Revd Benjamin Humphreys during the dramatic days of the great religious revival of 1905.

The village's second religion, rugby, was also closely connected with the local brew. Many famous matches were played on the recreation ground and then fiercely argued about afterwards in the Union Inn, once the headquarters of the Felinfoel club.

The old pub, a notorious bottleneck on the road through the village, was finally demolished in 1962 to widen the highway. It was said to be so close to the brewery opposite, that when the Union Inn's outbuildings were used by the company as a carpenters' shop, traffic ground to a halt when the men had to saw up planks.

Gradually the brewery built up trade throughout the old counties of Carmarthen, Cardigan and Pembroke, buying pubs as they became available. When David John retired from his business interests, his sons David and Martin took over the running of the brewery, with Llewellyn John looking after the Gorse tinplate works in nearby Dafen.

The brewery was registered as a private company in 1906, and in addition produced mineral waters under the Trebuan Spring label, the water being piped from a source above the village.

The family also had mining interests, and in 1908 these almost undermined the brewery. Sinking a well at the brewery to find an additional supply of water, workmen struck a two-foot thick seam of coal, some 12 yards below the surface. But, after due consideration, the John family decided not to work the seam, as it would interfere with the brewery buildings.

Another family joined in the running of the brewery when David John's daughter Mary Anne married John Lewis, the manager of the Wern Ironworks. This was later to lead to a serious split in the company, and the initial marriage was little happier.

John Lewis was a compulsive gambler, prepared to risk everything on the turn of a card or a throw of the dice. He is reputed to have lost a tinworks on a bet, and would probably have gambled away the brewery except his wife controlled the shares. He was little luckier in his business life, once buying a worked-out mine into which heaps of coal had been carted to fool potential purchasers.

In the 1920s the strain became too much and he shot himself while alone in the brewery office.

Undaunted, his wife Mary Anne, a formidable woman, carried on with the business. Her visits to the brewery made a deep impression on the staff. She carried a big stick and if she was unhappy with the performance of any of her employees – she hit them with it! The stick still hangs in the brewery office today.

GENTLEMAN'S DRINK: a stylish poster promoting
bottled Pale Ale in the 1920s

These were troubled times at the brewery, but the company's close connections with the tinplate industry were to alter all that.

Many types of meat, fruit and vegetables had been canned in the nineteenth century; the first food canning factory being established in London as early as 1812. By the end of the century the can was becoming a common sight on kitchen shelves. The 1895 edition of *Mrs Beeton's Book of Household Management* includes a picture of a huge array of tinned provisions including liquids like condensed milk and soup. But tinned beer was another can of worms entirely.

Many brewers were sceptical that drinkers would ever accept beer in a can. Customers expected beer to be on draught from a cask or in a glass bottle. Advocates of the can pointed out that ale had been enjoyed in pewter mugs for centuries, but few were convinced.

Much more of a hangover were the serious technical problems. Beer required a container that could withstand a pressure in excess of 80 lb per square inch. Food cans on the market only needed to withstand 25–35 lb. If filled with beer they would burst along the seam. At best, they leaked.

Then there was the question of flavour contamination. Beer reacted with the bare tinplate leaving a tinny taste. But coating with traditional brewers' pitch used in casks was no use in the smaller container. As a correspondent in the *Brewer and Wine Merchant* magazine explained: 'Samples of linings for the can were found to absorb all the hop flavour out of the beer and leave it tasting like the proverbial ditchwater.'

Finally there was the all-important bottom line. Cans cost more than glass bottles. So breweries, which had invested heavily in bottling plant and large stocks of returnable bottles, were unlikely to be enthusiastic.

In fact, like most drinkers, they were deeply suspicious. Sanders Watney of London brewers Watney, Combe, Reid, said in an article in the *World Press Review* in 1934:

I am not convinced that there would be any demand in this country for beer in cans. I cannot conceive the idea of a can ever replacing the half-pint, pint or quart bottle. The canning habit is certainly growing, but I do not think it will spread to drinks.

With brewers and drinkers indifferent, if not hostile, the impetus for change had to come from another quarter; the sector with the most to gain – the tinplate industry and the can manufacturers.

In 1909 the American Can Company (CanCo) had tried to produce a can for beer, but without success. Technical problems kicked the concept into touch. In 1931, anticipating the end of Prohibition in the United States and with the Depression affecting its conventional markets, it tried again.

It took two years of research to solve the pressure problem and to develop a half-satisfactory inner coating. Pleased with its efforts, CanCo then approached some of the major American breweries. They did not want to know.

The economics still did not add up, there was no obvious demand from drinkers and the larger breweries were wary of risking their reputations on an untried container. Two of the brewing giants, Anheuser-Busch and Pabst, had experimented with canning in 1929, and both had decided against pursuing what was seen as a poor, novelty product.

Eventually CanCo found a small brewery desperate enough to give tinned beer a try. The Gottfried Krueger Brewery of Newark, New Jersey, was in poor shape after thirteen years of Prohibition. To add to its problems, when the ban on alcohol was finally repealed in 1933, its workers went on strike.

As the can company was prepared to install the canning equipment for free – Krueger would only have to pay for it if the experiment was a success – the troubled brewery had nothing to lose.

In 1933 a test run of 2,000 cans was produced for a trial sampling. Further trials followed. The can was modified. Only in September 1934 did CanCo patent its Vinylite lining under the trademark 'Keglined'. Then in January 1935 two brands, Krueger's Finest Beer and Cream Ale, went on public sale in Richmond, Virginia.

The tinny took off beyond American Can's wildest dreams. Krueger was also pleased. Its sales shot up, so that by July its production was running at over five times its pre-canning level. By the end of the year, no less than thirty-seven US breweries were following its example and rattling out canned beer, including reluctant giants Pabst and Schlitz, brewers of 'the beer that made Milwaukee famous'.

A major factor in its success was that the packs were easier to carry home; shoppers were even prepared to pay a premium for the convenience. In addition, the compact can fitted more easily into the increasingly popular refrigerators appearing in every home; Americans preferred their beer ice cold. A survey of 750 drinkers found 89 per cent liked the new brightly designed containers.

To help convert wavering customers, early CanCo cans carried the imaginative claim that the contents were better than if served any other way, as the goodness was sealed in, the flavour preserved and the beer protected from the harmful effects of light.

American brewers discovered that while the unit costs of the one-trip cans were higher than returnable glass bottles, there were significant savings to be made in transport costs. Cans were much lighter and could be more tightly stacked – advantages which were also welcomed by retailers. And there were no empties to worry about.

Not surprisingly, the American glass industry was less enthusiastic. It started to kick the can. The Glass Bottle Blowers Association published a list of 'Fads and Fancies' – '1931 Miniature Golf; 1932 Flag Pole Sitting; 1933 Jigsaw Puzzles; 1934 Mah Jong; 1935 Tin Can Beer'.

In Britain, the main can manufacturer, Metal Box, had been watching developments in America with interest. But the company's leaders were far from convinced. 'I can't see the British public cottoning to the idea', wrote chairman F.N. Hepworth in August 1935.

Managing director Robert Barlow, the driving force of the firm, saw some possibilities: 'I can conceive of some of those beautiful establishments in Carlisle', he replied, referring to the modernized state-owned pubs in Cumberland, 'being rendered even more aesthetic by nicely decorated cans instead of ugly beer bottles.' But he added: 'The whole thing may come to nothing at all. I shall not risk any capital without your approval or without reasonable justification.'

If cans were to take off in Britain, the enthusiasm had to come from elsewhere. It did, from the bottom of the industrial pile – from the hard-pressed tinplate manufacturers of Llanelli in South Wales. They might be tucked away in a corner of Britain, but they had their eyes on the world.

Hard-hit by the Depression, they were desperate for more work. The manufacturers were in close contact with developments in the United States, holding regular talks by visitors and viewing industrial films from America. When canned beer exploded in the States, they seized the opportunity while others hesitated.

They were helped by having two brewers in the area who were just as eager to see increased employment and prosperity in Llanelli. Buckley's Brewery was reported by the Welsh morning newspaper, the *Western Mail*, to be investigating canned beer in October 1935. Felinfoel, with family interests in the tinplate industry, were just as keen – if not keener.

HISTORIC MOMENT: head brewer, Sidney John, celebrates the first canned beer in Europe with a drink straight off the production line in December 1935 (Felinfoel Brewery)

Robert Barlow of Metal Box felt there might be scope for canned beer in the export market. Anyway, he was not prepared to turn away the tinplate manufacturers and interested brewers. So an assembly line was set up at its Acton factory, with the inside of the can given an undercoating of lacquer and then a final lining of wax.

The John family's close connections with the tinplate industry were by now with the St David's Tinplate Works in Bynea. This was the company that supplied the sheets for the first British beer cans, assembled by Metal Box in London, before being returned to Llanelli for filling.

The weekly *Llanelli and County Guardian* recorded the historic moment on 3 December 1935, under the triple heading: 'Canned Beer Arrives', 'Epoch-Making Process at Felinfoel Brewery', 'New Hope for Tinplate Industry'. The paper's report said that the first can of beer was turned out 'without a hitch' in the presence of chairman Martin John, brewer Sidney John – 'who has pursued much research work in connection with the new idea' – and representatives of other brewing and trade interests.

The conical cans were filled on adapted bottling machinery and sealed with a standard bottle top (known as a crown cork). The 10 oz cans of pale ale were the equivalent of half-pint bottles. They were then packed in cardboard containers, holding two-dozen cans, ready for dispatch. The newspaper reported: 'One of the most impressive features of the process was its simplicity and speed. Girls, who in the past have handled many thousands of bottles, adapted themselves to new conditions with apparent ease and, once started, the cans were filled and corked with unbroken regularity.'

Brewery manager Willie Rees explained that Felinfoel had been experimenting with canned beer for a couple of months.

We were particularly struck with the success which followed its inception in America at the beginning of the year, and realising its potential and the stimulus which its universal adoption in this country would give to the tinplate trade, and especially local industry, we decided to put the novel idea into practical use ourselves.

Head brewer Sidney John even believed they had gone one or two steps better than the pioneers across the Atlantic, claiming the Americans had brewed a beer to suit the can, while Felinfoel had found a can 'to hold the perfect beer'. 'Their beer is being pasteurised and the result is that natural ingredients are being destroyed. That is not and will not be the case with our beer', he pledged. 'The difficulties of the London Metal Box Company have been to find a lining to preserve beer in its best state. After considerable research work, they have succeeded in doing so – the Americans have not.'

Felinfoel was so proud of its achievement that every employee of the brewery and the tinplate works was given a can to mark the occasion. These cans today are valuable collectors' items.

Buckley's was not amused at being beaten to the punch by its smaller neighbours. In the same issue of the paper, it paid for large adverts to proclaim:

The canning of beer was accomplished at Buckley's Brewery bottling stores on the 3rd December, and samples may be seen at the brewery and at displays in the town.

Until, however, the directors are satisfied that canned beer has the same estimable qualities as their bottled product, the process will be in the nature of an experiment, and for the time being the canning process will be conducted experimentally with persistence and caution.'

It was not until many years later that Buckley's actually produced a canned beer for public sale.

Felinfoel, in fact, would have welcomed Buckley's jumping on the can wagon, for the John family was more concerned to stimulate the tinplate industry than to corner a new beer market for itself. Llanelli was so depressed in the 1930s that Felinfoel even gave free bread and cheese in its pubs to customers buying its beer. 'Naturally, we are proud of what we have already been able to achieve', said manager Willie Rees. 'The success of the venture in this country, however, depends entirely upon the extent to which it will be exploited by the brewing trade.'

POLISHED PRODUCT: the first cone-shaped can of beer reminded many of a Brasso tin

Associated industries throughout South Wales were keen to see the product succeed and help lift the deep Depression. Sidney John estimated that 500 million cans a year would be needed if beer canning was generally adopted in Britain. That was big business, and some were prepared to go to extraordinary lengths to try and make it happen.

After trial batches Felinfoel began to produce canned beer for public sale from 19 March 1936. A month later the steel firm of Baldwins issued thousands of leaflets to its workers in South Wales urging them to buy their beer, cider, fruit juices and milk in tin containers instead of bottles, in order to encourage others to follow their example and so stimulate the tinplate trade.

'If the tinplate workpeople who have been accustomed to purchase bottles of beer will in future purchase cans of beer instead, they will start something that may create a large new volume of employment in South Wales', stated the leaflet. 'If they will not make such an effort to strengthen their local industry, they cannot expect people in other parts of the country to do so.'

That did not prevent industrial leaders trying to influence more distant areas of Britain. On 9 January 1936 some 3,500 miners after finishing their shifts at Sharlston Colliery near Wakefield in Yorkshire, came blinking to the surface to find a surprise: each man was presented with a gift of beer – in strange containers.

'The fact that the beer was in cans led to some speculation as to whether it was "the real stuff" or whether this was some misguided hoax. Curiosity gave way to satisfaction when the men had used the neat little opener provided with each tin', reported the *Yorkshire Observer*. 'The contents were sampled with much smacking of lips and expressions of appreciation of the donor's generosity.'

Each can of beer was accompanied by a card wishing each miner 'A Happy New Year'. The donor was Sir William Firth, chairman of Richard Thomas & Co. of South Wales, the parent company of the Sharlston Colliery. The canned beer was American, but Sir William said he hoped to supply British ale in similar form next year.

Certainly there would be no shortage of beers to choose from. After the lead of Felinfoel came Jeffreys of Edinburgh with a canned lager. Barclay, Perkins and Hammertons of London, Simonds of Reading and McEwans and Tennents in Scotland soon followed.

Metal Box began to wake up to the potential, and in May 1936 managing director Robert Barlow invited journalists and other interested parties to inspect its Acton works. In July it placed an advert in the brewing trade press proclaiming, 'It's arrived – eleven breweries have now ordered beer cans!' The saving in weight against the bottle was said to be 59 per cent; in space 66 per cent.

Metal Box also produced large showcards (made, of course, from tin) to promote the new concept in pubs and off-licences. Against the background of a sports crowd, the wall display spelt out the claimed advantages of 'Beer in Cans' – hygienic (used once only); slips into the pocket; no deposits; no returns; unbreakable; protected from light; opens like a bottle; easier to pour; handy in the home; a modern idea. 'And it's good! Beer canned for your convenience. Try it', urged the placard, concluding, 'Why didn't they think of this before?'

Felinfoel Brewery workers even became film stars, revealed the *Llanelli and County Guardian* of 9 July 1936:

The growth of Llanelli's new industry – canned beer – brought further recognition to the town on Monday when a newsreel cameraman, representative of one of the biggest film companies in the world, presented himself at the Felinfoel Brewery and obtained the permission of the management to shoot a film of the processes.

And so for a day the long-established brewery took on the temporary garb of a film studio as mysterious gadgets were erected in the various departments. High-powered lights and glistening screens were thrown up here, there and everywhere, and for a few brief hours the female workers fulfilled their life's ambition – to figure on the modern film screen

The feverish activity continued after the cameras rolled away, as more than a quarter-of-a-million cans were consumed in the first few months. Felinfoel took on extra workers with 'a distinct likelihood' of more jobs to follow. A consignment was even sent on a world cruise to see how canned beer stood up to the tropics.

By the end of 1936 two million cans had been produced, and by October 1937 some twenty-three British breweries were marketing over forty different brands. Among them was Lassell & Sharman of Caergwrle, in North Wales, which canned its Famous Strong Ale, while Felinfoel marked the coronation of King George VI in May 1937 with a special canned ale.

ROYAL CELEBRATION: the brewhouse decked out for King George VI's Coronation in 1937, which Felinfoel marked with a special canned beer (see back cover)

But despite this flurry of activity, canned booze was still small beer. Many of the factors that triggered the explosion in America were less significant in Britain. Transport savings, which could mount up in the wide-open spaces of the United States, counted for less in the much smaller British Isles. Take-home beer was not so important in pub-orientated Britain, and when they did want a smaller container, Britain's conservative drinkers still preferred glass bottles.

As the *Brewers' Journal* rather bluntly noted a year after Felinfoel had canned the first beer in Britain: 'It is rather a handicap to the new idea of canning beer that the present type of container suggests immediately to the consumer that he is being regaled with metal polish' (the cone-top can looked like a Brasso tin). 'This creates a destructively critical atmosphere towards a really sound proposition,' continued the article, 'and one hears more remarks about the containers than the contents.' Drinkers were certainly not prepared to pay more for these odd-looking vessels. Felinfoel did not replace them with the more familiar flat-top cans until after the Second World War.

An influential article in the *Financial Times* of 23 January 1936 spelt out the figures:

The quotations given (for cans in large quantities) have finally demonstrated that the can is an uneconomic proposition in the home market. A half-pint can, overprinted attractively enough to rival its sister bottle, would cost approximately 12s. per gross. Bottles of the same size cost no more, and are good for an average of 200 uses.

There were also still quality problems. Metal Box in its official history by W.J. Reader admitted that 'the wax lining then used spoilt the taste of the beer'. When Robert Barlow had taken his party of visitors round the Acton factory, they had been offered a canned light ale which, according to the *Brewing Trade Review*, 'was not quite star bright'. W.J. Reader concluded that 'by 1939 these difficulties were still unresolved when the whole experiment was overwhelmed by the outbreak of war'.

Canned beer in Britain only scored heavily on one front in its early years – exports. Here the beer was being shipped long distances and so the savings in weight, space and freight charges were significant. And there was no need to worry about getting the bottles back from Singapore.

With the war canned beer production was almost completely usurped for the armed forces overseas. The Americans even camouflaged their cans, painting them in green varnish known as 'olive drab', in order not to attract the attention of enemy snipers. Felinfoel, because of its tinplate industry connections, was the only British brewery allowed to continue selling cans for home consumption. But most batches went to the distant war fronts, shipped out by the NAAFI.

One of the few cargoes that managed to break the German siege of Malta contained cans of Felinfoel for the parched Llanelli Territorials manning the Mediterranean island's anti-aircraft guns. Thirsty 'desert rats' in North Africa also had good reason to thank the brewery for a little canned laughter. Thousands of cans reached troops in the Far East.

After the war Felinfoel lost its armed services contracts, but the distant contacts were not entirely forgotten. In 1976, thirty years later, Terry Beynon, a native of Port Talbot, visited the brewery to ask for some beer – for his Peacock Bar in Penang, Malaysia. In January 1977 the first 100 dozen cans of Double Dragon were sent out. Terry Beynon wrote after the first shipment, 'It is very popular with the locals. We ran out of your beer before the second consignment left the UK.'

The success rekindled Felinfoel's enthusiasm for overseas markets, becoming the only South Wales brewery with a regular interest outside Britain. Later many beers were brewed expressly for export, like St David's Porter, Prince's Porter, Cream Stout, Heritage Ale and Hercules Strong Ale, with California becoming a major market in the 1980s, accounting for some 650 barrels a year.

But these beers were sold in bottle. Canning did not prove to have a silver lining for Felinfoel, since the market was rapidly dominated by the major brewers. In fact, in the immediate post-war period the brewery struggled to survive.

Fred Cheesewright, who was head brewer from 1951 to 1982, claimed that when he arrived the brewery was on the brink of collapse. 'When I first came here, Felinfoel had a terrible name. In one year the brewery lost 30 per cent of its trade. Some houses were selling as little as a kil [18 gallons] a week. We were close to bankruptcy.' It took a long time to turn the brewery round, as the plant and the seventy to eighty pubs had been neglected for some years, the war starving the company of investment.

The sorry situation was not helped by a bitter disagreement between the John and Lewis families, which came to a head in 1965 when larger neighbours Buckley's bid for the brewery.

The Lewis family had been in charge for a number of years, with most of the members on the board, but it was a distant control. After the Second World War the brewery's head office had been moved to Knightsbridge in London, where the Lewis family ran other interests, including a motor company. The John family decided to sell out.

Buckley's had privately approached members of the John family before making its £500,000 bid public. So when the attempted takeover first became news in April 1965, they could claim to have secured acceptance from 48.7 per cent of shareholders. By May this figure had edged up to 49.5 per cent. Buckley's was almost there. 'We have been motivated solely by our determination to ensure that the brewing industry in Llanelli remains under local control', said Buckley's, adding, 'With the interest that Buckleys now have, this object should be achieved.'

One Hundred Years — And Going Strong

T hat was the message from the Felinfoel Brewery in 1978 when the company celebrated the centenary of the old brewhouse, built in 1878, with one of the strongest beers ever brewed in Wales.

Head brewer Fred Cheesewright spent months experimenting before finally settling on a brew with staggering potential. With an original gravity of 1100, a half-pint had the strength of at least 2½ whiskies. Altogether 48,000 half-pint bottles of Centenary Ale were produced during the year.

'It is not too sweet, which is one of the faults that can be attributed to high-gravity beer, and from the first drink one can appreciate there is a punch in the bottle', said Fred Cheesewright, who is pictured carefully sampling the knockout drop with general manager Peter Donaldson (left).

Felinfoel's manager Cyril Marks retorted: 'We have no intention whatsoever of being taken over by anybody, local or national.' Felinfoel's directors recommended rejection – 'or it could mean the closure of the brewery and our employees losing their jobs'.

A crucial role was played by Lady Davies who owned eleven shares. She was approached by Buckley's and offered £2,750. If she had sold out, the takeover would have succeeded. Instead she rejected the money and gave the shares to the Lewises.

The family remained firm. Chairman Trevor Lewis said it held the rest of the shares and had no intention of selling. Despite its paper-thin majority, the family kept control, though

Buckley's gained a seat on the board. A holding company was established to prevent further family disputes leading to more shares, and a controlling interest, sliding down the road to Llanelli.

The attempted takeover left one strange ritual. As Buckley's had contracted with members of the John family to buy their shares, it ended up holding 49.5 per cent of Felinfoel without any influence in the private company. Buckley's shareholders were not amused and regularly asked at AGMs when the takeover was going to be completed. So Buckley's made frequent offers for Felinfoel, to satisfy their shareholders' demands, which the Felinfoel directors just as regularly threw in the bin.

In the 1970s the company under Trevor Lewis's son John began gradually to modernize the old brewery. The wooden fermenting squares were replaced with stainless steel ones, and a new copper was installed in place of the original, heavily-patched, open coal-fired vessel. 'We don't have to work in a continuous fog any more', said head brewer Fred Cheesewright.

As a sign of renewed faith in the village brewery, when Trevor Lewis died in 1974 the head office was brought back to Felinfoel.

The hard work was rewarded in 1976 when its premium bitter, Double Dragon, won the brewing world's equivalent of the world cup, the Challenge Cup, for the best cask beer at the Brewers' Exhibition in London, held once every four years. Its standard bitter also took first prize in its class. General manager Peter Donaldson described the double victory as a 'tremendous achievement', particularly for Fred Cheesewright. 'He has contributed in large measure to our success. In fact, since he arrived on the scene 25 years ago, there has been a very marked improvement in our trade.' By now the company was brewing some 400 barrels a week.

To celebrate, Felinfoel named a pub in Morriston the Champion Brewer. It was a fitting name for a brewery that had survived and then triumphed again in the modern era.

THOSE WERE THE DRAYS: a line-up of steam and petrol lorries in the 1920s (Felinfoel Brewery)

THE BURTON OF WALES

From time immemorial the ales of Wrexham have been held in high estimation, their superior quality being due to the celebrated wells with which the place abounds; these wells supply the finest water for brewing and domestic purposes. Indeed, Wrexham may now be called the Burton of Wales.

Alfred Barnard, *The Noted Breweries of Great Britain and Ireland*, 1892

Brewing was a significant industry in most of the major towns of South Wales, but only in one urban area did the lucrative liquid trade in malt and hops rise up to dominate all other businesses.

Like Burton-on-Trent in England, Wrexham in North Wales became a brewing centre, known far and wide for its ales. In Cromwell's time the home-brewed beer was already so popular that soldiers in the Civil War were said to have deserted his army in Chester and crossed the border to enjoy a good drop.

This rash and risky move becomes even more understandable when one recalls the reputation of Chester beer. Sion Tudor was driven to describe it in verse:

> Chester ale. Chester ale! I could ne'er get it down,
> Tis made of ground-ivy, of dirt and of bran,
> Tis as thick as a river below a huge town!
> Tis not lap for a dog, far less drink for a man.

Famous magazine columnist Nimrod (Charles James Apperley of Plas Grono) wrote in praise of Wrexham beer in 1842: 'And what excellent ale it was! I have drunk nothing like it since. And I have reason to believe that Wrexham is still celebrated for the excellence of its home-brewed ale.'

When author and adventurer George Borrow passed through in 1854, during his famous tour of *Wild Wales*, he asked a group of idlers by the imposing parish church of St Giles if they spoke Welsh?

'No, sir,' said the man, 'all the Welsh that any of us know, or indeed wish to know, is Cwrw da.' Noted Borrow: 'Here there was a general laugh. Cwrw da signifies good ale . . . I was subsequently told that all the people of Wrexham are fond of good ale.'

A visitor in 1860 reported:

During my stay I have not tasted a mediocre glass, but that which excelled all others was the beverage brewed by Messrs T. Rowland of the Nag's Head Brewery. We smacked our lips in ecstasy. Two or more glasses made us feel quite patriotic and in good humour with everybody, and I shall take pride in extolling the virtues of Wrexham Ale.

The quality of the beer was even blamed for making the stage-coaches run late. An inquiry into why the Chester to Shrewsbury coach took double the time to cover the distance than other similar-length routes discovered that the driver stopped in Wrexham to allow passengers a refreshing glass or two of the local ale – and then had great difficulty in getting them to complete their journey.

The passengers were certainly spoilt for choice. As early as 1835 Pigot's Directory is recording that 'the malting trade is of consequence, and there are several respectable breweries'. By the late 1860s there were nineteen commercial breweries in operation. Cardiff was to have as many, but brewing was well down the list of major industries, being noted in directories at the tail-end along with pickle manufacturers and dog biscuit makers. In Wrexham, as Porter's Directory of 1886 records, 'brewing forms the staple trade of the town'.

Worrall's Directory of 1874 even names names in its introduction to the town:

Wrexham has always been celebrated for the fine quality of its ale, and consequently the brewing trade has flourished here. Within the last few years, however, this trade has enormously increased, and several very spacious and handsome breweries have

been built at great cost. The principal of these are the breweries of Mr Peter Walker, Messrs Joseph Clark and Sons and Mr John Jones (Island Green Brewery).

The brewers and innkeepers ruled the town. When a preacher spoke against the evils of drink, he was run out by the authorities. During the first fifty years after its incorporation as a borough in 1857, brewers made up well over a third of Wrexham's mayors. Even the old Town Hall was converted into a bonded wine and spirits store, with an elaborate bar.

Business was helped by the fact that the bustling market town was home to a large army garrison. One Victorian wag claimed: 'Wrexham beer is made from mashed sheet

CIVIC PRIDE: even the old Wrexham Town Hall was turned into a bonded store, run by Thomas Williams

music and boxing gloves, for it makes one either sing or fight.' Reverend David Howell, giving evidence before the Lords Select Committee on Sunday Closing in 1880, said: 'Wrexham has gained more notoriety, for being fond of drink, than any other town around.'

The police stated that 'drunkenness was the worst in Wales'. Wrexham was known as 'the most fractious town in Britain'. William George, brother of Lord Dwyfor, claimed: 'Beer and baccy are the creed of the town – there are too many public houses.' An 1849 Board of Health inquiry found that for a population of 7,000 'there is an unreasonable number of innkeepers and publicans' – in total sixty pubs, five beer shops, four spirit vaults and twenty off-licences.

Wrexham had a public brewhouse in Brook Street at the bottom of Town Hill, where townsfolk could take along their own malt and hops and use the brewing equipment for a small fee. This public 'brewing kitchen' was mentioned in 1700 and was sold a century later with a house and stable.

The reason why Wrexham became so famous – and notorious – for its beer was simple. Like Burton-on-Trent, it had ideal water for brewing. The Brynyffynnon Spring in the lower part of the town was well known for its medicinal value, having a high mineral content, particularly in sodium and calcium sulphate. All the ale breweries were packed around this area, to the east of the Wrexham fault where most of the town lies. Only one brewery drew water west of the fault, and that was the Wrexham Lager Brewery which required a softer, different type of brewing water.

There was a second reason for this concentration of breweries. The town was historically divided into two manors, Wrexham Regis and Wrexham Abbot. Those brewing in the royal manor had to grind their malt at the King's Mill – and pay a toll for the privilege. Those brewing in Wrexham Abbot avoided this tax. Not surprisingly, therefore, brewers flocked to the Abbot.

The first commercial brewery, as distinct from the many attached to public houses, appears to have been established by Edward Thomas in 1799 in College Street. He converted an old tannery off Town Hill into a brewhouse, but died soon afterwards. The business was taken over by Edward Crewe, later becoming the Albion Brewery run by John Beirne (Mayor of Wrexham 1876–7) of Plas Derwen.

The confidence and pride of the town's brewers in the quality of their water and their ales is best reflected in a tongue-in-cheek opinion piece penned for the Cambrian Brewery of Bridge Street in 1892 for the promotional book *Wrexham Illustrated*. The brewery owner, William Sisson, enjoyed a good dig at English plans to pipe water out of Wales.

It is not to be wondered at that London and Birmingham should be rivals for the possession of some of the Welsh water. We have some experience, however, of these two large cities, and our impression is that their object in trying to annex the water supply of Wales is not so much owing to their love for pure water in its natural state as to their desire to procure it for manufacturing purposes, so as to produce for themselves beers and other beverages that should resemble those produced in Wales itself.

The pretended love for pure drinking water on the part of the enterprising citizens of London or Birmingham is somewhat amusing, as no natural-born inhabitant of either

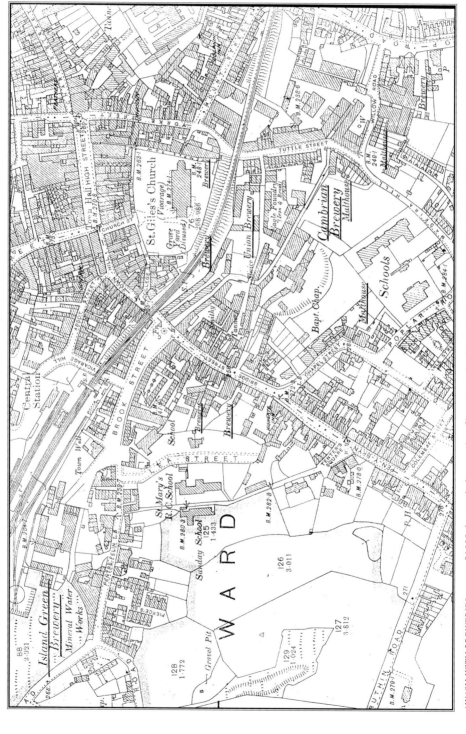

AWASH WITH BREWERIES: a map of 1899 showing the huge number of breweries and malthouses in the centre of Wrexham (Ordnance Survey)

town ever tasted pure water in his life, or would appreciate it if he did. The true secret of the matter is that some adventurers from these towns, noting the exquisite flavour and quality of the Wrexham ales and stout, and having been told that the Penadur and other Welsh springs were sources of marvellously fine water, conceived the idea of tapping the Welsh water supply and making similar ales themselves. At least, that is our private opinion, though we give it freely, without reserving the copyright.

After judiciously sampling the ales of the Cambrian Brewery, we had no doubt on the subject; but we do doubt whether, even with the water laid on, such delicious beverages could be brewed in Birmingham or any other district. The fact is there is a special 'genius loci' which gives to the noted Wrexham ales their peculiar excellence, and it would be vain to think of reproducing these qualities elsewhere.

William Sisson had taken over the Cambrian Brewery in 1874 from Joseph Clark & Sons, completely remodelling the three-storey brewhouse with a modern 10-quarter plant to produce 'a large stock of choice strong and mild ales and stouts'. In 1892 Sisson employed some twenty men. His proud emblem was the three feathers of the Prince of Wales.

Most Wrexham breweries were of similar if not smaller size. Charles Bates' Union Brewery in Tuttle Street, founded in 1840, was described in *Wrexham Illustrated* as 'a compact, well-arranged structure', adding that 'quality is the main object here rather than quantity'. The cellars provided storage for about 600 casks of pale ale and stout.

When Julius Chadwick bought the Burton Brewery in Bridge Street in 1875 for £541 from Mrs Mary Evans, wife of the late tenant Charles Evans, he received the following plant, according to the inventory. In the brewhouse there was a wrought-iron brewing pan and brass tap, cast iron pan, tinwork pump, ale cooler, stage, hopper and steps, under vat lined with lead, piping, mash tub, two stillages, four tubs, two hogsheads, a rake, two shovels, a brush and a gauge.

In the 'tunning room' were 6 puncheons, 10 working tubs, 5 tin cans and 3 funnels. In the cellar were 25 barrels (36-gallon casks), 96 kils (18-gallon casks) and 6 pins (9-gallon casks). There was also a malthouse and an engine room. This was a small-scale brewing business and seventeen years later not a great deal had changed. 'Although it cannot be said to rival the gigantic concerns which boast of their connection

COMPETING CONCERNS: bottle labels from two of Wrexham's many breweries (Keith Osborne)

with Burton-on-Trent, still it is capable of turning out a fair quantity of goods, while quality is made a very particular point', claimed the company. The three-storey, red-brick brewhouse could mash 'four quarters of malt at a time'.

The problem for these modest family businesses was that they were on tight town-centre sites, packed close together with little space to expand. Even more pressing was the fact that they were all competing for trade in a medium-sized town. Bills show that Thomas Openshaw, licensee of the Golden Lion in Wrexham in the early 1890s, was taking beer from the Burton Brewery, FW Soames, John Jones' Island Green Brewery and the Wrexham Lager Brewery. The population of Wrexham in 1891 was only 12,552, though there were many collieries, steel and brick works in the immediate neighbourhood. Beyond this surrounding industrial belt, they were chasing low-volume business through the scattered rural villages and small towns of North Wales. It was not a recipe for easy success.

A number of the firms repeatedly changed hands in rapid succession. Some went bankrupt and a few companies tried to develop links with Merseyside in order to supply the larger trade there. Some specialized. Chadwick's Burton Brewery took special pride in its 'celebrated Welsh stout' (*y stowt cymreig pur*) – 'which is regarded in the neighbourhood as quite a national beverage'. The company was also one of the chief bottlers of Guinness and Burton ales in the town.

Thompson's Sun and Eagle Breweries concentrated on their strong Welsh ales, sold as far afield as Cardiff under the slogan 'Cwrw Da Am Byth' (Good Beer For Ever).

'The Wrexham strong ales have gained a reputation that is not confined to North Wales. These beverages form a leading line with Thompson and Co., and have been associated with their brewery for a very long time,' said *Wrexham Illustrated*. 'The specialities in strong ales are certainly remarkable beverages, full-bodied, strong and fine-flavoured.'

The Sun Brewery in Abbot Street was described as having 'a quaint picturesque appearance'. The business claimed to be 'probably the oldest in Wrexham', having evolved in the 1790s from the brewhouse attached to the Sun Inn. Its principal cellars were constructed out of a subterranean passage connecting Hope and Abbot Streets. Adverts proclaimed: 'Ask for the Little Brown Jug of Ale, and see that it is Thompson & Co.'s.'

CWRW.DA: the trade mark of Thompson's Sun and Eagle Breweries, emphasizing the company's Welsh character. Thompson's specialized in strong Welsh ales and, judging by this sketch, these were offered in special narrow glasses

By 1892 the Eagle Brewery in Bridge Street was only used for stores by Thompson's, the previous owner, Robert Williams, having also specialized in strong Welsh ales. A price list shows him producing four grades in

1880, ranging from 80s. to 60s. a barrel. Thompson's was in addition the local agent for Watkin's London Stout and produced mineral waters 'made from the celebrated Brynyffynnon Spring, which is noted for its purity and excellent keeping quality, that no other manufacturers in the neighbourhood can surpass us'.

John Beirne combined the Albion Brewery with his second business as a tallow chandler (candle-maker). John Murless, twice Mayor of Wrexham, ran the Wynnstay Brewery as a sideline alongside the grand Wynnstay Hotel in the High Street, with stabling for 200 horses. He also acted as agent for the Great Western Railway, looked after the station refreshment rooms and traded as a wine and spirit merchant from the Wynnstay Old Vaults in Yorke Street.

Only three ale breweries in Wrexham managed to break through this restricted scale of brewing. Two firms – Soames' Wrexham Brewery and the Island Green Brewery – combined to form Border Brewery, which outlasted all its local rivals. The other was the largest of them all and could have established the town permanently as a major brewing centre, but petty jealousy and thwarted ambition cut off its progress in its prime.

The name Peter Walker is one to conjure with in brewing circles. His name is still on drinkers' lips in Liverpool, where the giant Carlsberg-Tetley combine operate a chain of traditional Peter Walker pubs. Tetley brew at the huge Walker Brewery in Warrington to this day, producing a range of Walker brews alongside other products.

The Walkers were men of vigour and vision – and powerful personalities. In 1837 Peter Walker, the son of a wealthy colliery owner, came to Liverpool from Scotland, where he had run a small brewery in Ayr. Soon he was running a more successful brewery on Merseyside, with an increasing number of pubs and associated wine and spirits businesses, helped by his two sons Peter and Andrew.

Andrew rapidly became the dominant figure, taking over a second brewery in Warrington in the late 1840s and then building another huge plant in the Cheshire town after 1866. By the 1890s these giant premises in Dallam Lane had a capacity of 10,000 barrels a week. He pressed ahead relentlessly with his ale empire, buying up more and more pubs. Depots were established in Birkenhead, Chester, Crewe, Manchester, Leeds, Newcastle, Belfast and Dublin. It is said that by the early 1890s his company owned more tied houses than any other brewery in Britain.

Sir Andrew, as he became, made many grand charitable gestures, such as the Walker Art Gallery in Liverpool, opened in 1877, and the Walker Engineering Laboratories in 1889. In his later years he lived in well-earned splendour, his conspicuous extravagances including country estates, steam yachts and elaborate entertainments. On his death in 1893 he left a vast personal fortune of £2.9 million.

No one ever dared stand in his way. His father had retired back to Scotland in 1873. Andrew's brother, Peter Walker junior, escaped his immense shadow earlier, first coming to Wales when little more than a youth. He could never hope to emulate his hugely successful brother – but he made a brave and determined attempt.

Peter Walker arrived in Wrexham in the late 1830s as a pupil brewer to Joseph Clark at Clark and Orford's Cambrian Brewery in Bridge Street, later superintending building work there. Though he returned to Liverpool to develop several wine and spirit businesses, he retained his links with the Welsh town, living at Coed-y-Glyn, near the entrance to Erddig Park, from the mid-1840s.

Experienced in the trade through working with his family, he launched out on his own in 1860, buying a small brewery in Willow Road, Wrexham, from Richard Evans. Slightly away from the cramped town centre, this site provided room for expansion and he extended the premises considerably to create the largest brewery in Wrexham. Some of his sales were in Liverpool where he retained offices and stores in Argyle Street.

His rapid rise to eminence in Wrexham is sharply illustrated by the fact that he was created mayor in 1866–7, at the age of forty-six, two years after his mentor Joseph Clark had held the office. He was elected mayor again the following year, when he provided a new pulpit for Wrexham Parish Church. Peter Walker also gave Wrexham Corporation its mace. He had become a wealthy man of substance in his own right and a leading local figure.

Further extensions were made to the Willow Brewery in 1871, a commemorative stone marking their opening. The buildings dominated the area, the 140 ft tall chimney towering over Wrexham. In 1877 he rebuilt Salop Road Bridge near the brewery at his own expense. The company's annual dinner was a prime event in the social calendar, housed in the brewery's own banqueting hall.

But, so the story goes, Peter Walker was unhappy with his adopted town. He believed he deserved to be mayor again, but the honour of a third term escaped him. Instead, his former close brewing rival, Thomas Rowland of the Nag's Head Brewery, was made mayor for the second time in 1881. Peter Walker resolved to quit Wrexham and take his brewing business to the chief capital of ale, Burton-on-Trent.

The decision was a blow to the Welsh town and its reputation as a major brewing centre. The Willow Brewery was a substantial loss, closing in September 1883. The extensive premises – 7,000 sq. yd in size – were later bought by Wrexham Corporation and were able to comfortably house a central council depot with workshops, stables and offices, refuse destructor, electricity generator, public baths, a gymnasium, public laundry and assembly rooms.

Perhaps even more of a loss was the brewery that might have been built in Wrexham, for once Peter Walker had decided to move lock, stock and barrel to Burton, he began planning a grand 50-quarter brewery there on a 10-acre site alongside the Midland Railway, with a siding running into the premises. The *Brewers' Journal* eloquently described the palatial pile in Clarence Street, built with many extravagant flourishes in November 1882:

> The building is substantially built in red brick with stone cornice, mouldings and string courses. The panels in the walls under the windows are filled in with Brown's patent tiles, specially designed and ornamented with hops and foliage. The roofs are framed in iron and covered with Welsh slates, the brewery house portion having glazed dormers and lantern, surmounted by ornamental iron cresting and finials.

The towering maltings alongside had a distinctive octagonal kiln, topped by a Welsh goat-shaped weather vane, which can still be seen swinging in the wind today.

There was an underlying reason why Peter Walker was creating such a magnificent monument to the art of brewing in Burton-on-Trent. Five years before, his brother Andrew had also built a brewery in Burton to brew pale ale, but it was a modest affair off the Shobnall Road with no maltings. For once, Peter could upstage his more eminent relative.

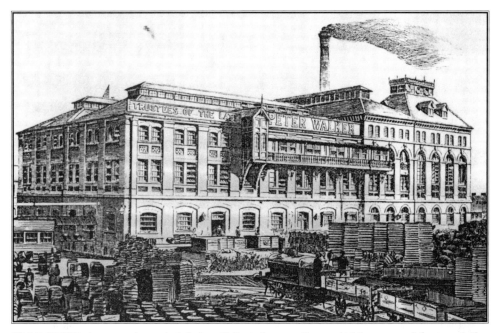

GRAND DESIGN: the imposing brewery built for Peter Walker in Burton-on-Trent — which he never saw. Before it was half completed, he had died

Sadly he never saw his crowning glory. The foundation stone was laid on 17 February, 1882, but he died at Coed-y-Glyn less than two months later on 13 April, aged sixty-two, before the building work was even half finished. His trustees completed the construction and, somewhat incongruously, plastered across the soaring architecture the large words 'Trustees of the late Peter Walker'.

His ambitions had been cruelly cut short. He was also to have been the Conservative candidate for Denbighshire at the next election. He left a considerable estate of £227,000, with ample provision for each of his three daughters.

If the people of Wrexham were unhappy that he was moving his brewing business out of the town, it never showed at his funeral. The procession was said to have been 'one of the largest ever seen in Wrexham'. He did not forget them. His trustees honoured his pledge to give £1,000 to the new national school on Madeira Hill, above the Willow Brewery.

Ironically, his brother's company Peter Walker & Son, registered in 1890, brought the famous name back to Wrexham in 1909 when the Warrington brewers bought Charles Bates' Union Brewery in Tuttle Street, close to the old Willow Brewery. The company even used Peter Walker's old maltings in Willow Road. It was a short-lived revival; in 1927 the company sold off its twenty-seven pubs in the area, the Island Green Brewery taking thirteen of them.

All that remained to remind locals of the grand brewer was his Willow Brewery premises – until they were severely damaged by fire in the early 1970s and demolished. The chimney, which had dominated the skyline for a century, was dismantled in 1973.

NAG'S HEAD IN FRONT

With the demise of Peter Walker's brewery, the standing of Wrexham as a significant centre of ale brewing rested on its two remaining companies of size – the Nag's Head Brewery and the Island Green Brewery. It was a shaky and uncertain foundation.

The Nag's Head Brewery grew out of the pub of that name in Mount Street, dating from the mid-eighteenth century, which developed an early reputation for the excellence of its ales. 'A visitor who had passed through Wrexham without sampling the home-brewed of the Nag's Head would have been regarded as having failed in his principal and most obvious duty, and as a very eccentric person in all respects', claimed a promotional portrait.

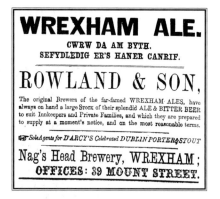

WREXHAM ALE.

CWRW DA AM BYTH.
SEFYDLEDIG ER'S HANER CANRIF.

ROWLAND & SON,

The original Brewers of the far-famed WREXHAM ALES, have always on hand a large Stock of their splendid ALE & BITTER BEER to suit Innkeepers and Private Families, and which they are prepared to supply at a moment's notice, and on the most reasonable terms.

☞ *Sole Agents for* D'ARCY'S *Celebrated* DUBLIN PORTER & STOUT

Nag's Head Brewery, WREXHAM;
OFFICES: 39 MOUNT STREET.

BOLD CLAIM: *William Rowland advertised that he and his son were 'the original brewers of the far-famed Wrexham Ales'*

An advert of 1880 said its ales had been 'celebrated for one century', but it did not develop as a commercial brewery until after William Rowland bought the old inn in 1834 and expanded the premises with the help of his son, Thomas, who became Mayor of Wrexham in 1868.

In the early 1870s the business (valued at £46,300) was bought by Henry Aspinall, an ambitious brewer from Birkenhead. He made waves from the start. He formed the Wrexham Brewery Company in 1874 with local wine merchant William Overton and was soon able to boast about his 'prize-medal ales' as his beer took first prize at the Royal Albert Hall Exhibition in London in 1875. Stores were opened in Rhyl and Shrewsbury.

In a bid to gain more land to enlarge the brewery, he made an offer for the roadway and ancient space below the neighbouring parish church. Wrexham Corporation agreed – but not all townsfolk were happy. Some believed that the open space was common land, belonging to the public since time immemorial, and that the area formed one of the most attractive approaches to the town centre, with its view of the elegant church with its grand Gothic tower. Many muttered that the brewers had too much influence in the town.

A few must have been quietly pleased, therefore, when Aspinall's plans collapsed around his head. At the instigation of a creditor, he was made bankrupt in 1879 owing the 'very considerable' sum of £50,000. Thirteen years later his case was still troubling the Wrexham County Court. The Wrexham Brewery Company went into liquidation.

Despite Aspinall's financial difficulties, Wrexham's reputation as a brewing centre was still sound – certainly high enough to attract a brewer all the way from the far side of England. Arthur Soames from the malting town of Newark in Nottinghamshire bought the brewery in 1879, making his 21-year-old son Frederick the manager.

The young man put all his energy into the venture, trebling the business in less than ten years. During the 1880s he enlarged the brewery considerably, building a new 50-quarter brewhouse making the company the largest brewer in Wrexham. The firm traded as F.W. Soames, with a bridled horse's head appropriately as its symbol.

When Alfred Barnard visited Wrexham in 1892 as part of his research for his work, *Noted Breweries of Great Britain and Ireland*, he inevitably headed straight for Soames. No other brewery was mentioned. 'We were shown up to Mr Soames' private office, strewn with papers, showing the immense amount of work to be done in a business of this size', wrote Barnard.

We were particularly interested in a statement from Mr Soames, to the effect that when he commenced the business he determined to brew a light thoroughly finished beer, characteristic in every way of the true Welsh ale, as opposed to the common heady ale, since he was convinced that the taste of the general public would improve, and also because such ale was bound to assist the cause of real temperance.

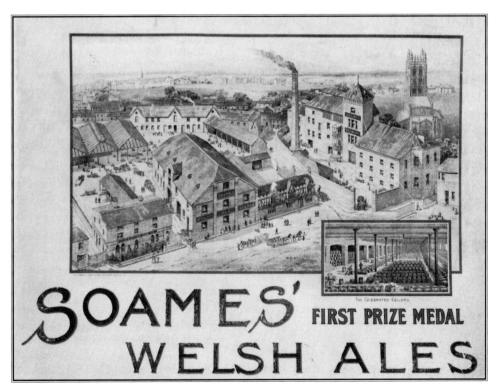

WREXHAM'S BEST: Soames' brewery in the 1890s with the half-timbered Nag's Head pub in the foreground. A somewhat reduced parish church is tucked behind the four-storey brewhouse (Clwyd Record Office)

Frederick Soames had learned fast about the politics of his adopted land. Only a Welsh brewer would be at pains to claim he was brewing a beer 'to assist the cause of real temperance'. Soames added that the colliers 'particularly approved of this finer quality of ale'.

Alfred Barnard was shown round by the head brewer, Mr G. Ross, who had trained in Burton-on-Trent and then been head brewer for eight years at the Trent Valley Brewery in Lichfield. Barnard was suitably impressed by what he saw. 'The new brewhouse, situated in the west yard, is a lofty, four-storied, red brick building which, with its ornate hoist cages and square tower, presents a fine appearance.' Over eighty men were employed on the premises.

He was intrigued by the fermenting house, with its nine 80-barrel vessels, some made from pine and others from slate. But not too fascinated to be taken unawares. 'A visitor is generally asked to hold his head over these fermenting tuns, to smell their contents; but we were not so to be caught.' He did not intend to be knocked out by the rising carbon-dioxide from the fermenting beer. An 'enormous' settling-back chamber held a further ten slate squares. He noted:

> In this brewery immense importance is attached to the fermentations, and to the effect of this portion of the manufacture upon the quality, condition and character of the finished ale; and as Messrs Soames produce strong and mild ales, pale ales and stouts, we were interested to learn that three systems of fermentations are carried on, and that two distinct types of yeast are used.

> Below the floor is the deep well chamber, lined with an immense iron curb, having a perforated bottom, through which penetrates the water, clear, pure and sparkling; the depth of the spring from which it flows preventing all possibility of impurity. This water is pronounced by chemists to be very similar to the Burton brewing water.

In the east yard, on the other side of Tuttle Street that ran through the middle of the brewery, the visitors examined the new three-storey ale stores, which were fitted with an hydraulic lift. Here in a 160 ft x 42 ft building the beer was racked into casks or bottled. 'The basement is entirely devoted to beer cellars, where additional cellarage has been obtained by excavating beneath the yard, the roof being supported on stout iron columns.' The cellars covered an area of 10,000 sq. ft, capable of holding 5,000 casks. 'We were somewhat taken aback at seeing such spacious and extensive cellars', admitted Barnard.

The bottling department was also a surprise:

> What most struck us was the store-room. We have never seen such a light, bright and clean apartment used for bottled beers. On the floor, ranged in symmetrically built-up stacks, are hundreds of dozens of bottles placed upright, with their smart labels turned to the beholder, looking like pyramids of glass sandwiches. Each layer of so many dozen has a board placed between it, and so firm are they built that half-a-dozen persons might stand upon them without any danger.

No brewery visit is ever complete without a sample or three. 'We tasted two or three specimens of Messrs Soames' brew and were specially taken with their light bitter ale, so clean, bright, light and full of life and condition.' He was informed that this beer, called the Guinea Wrexham, was a speciality of the company, being 'immensely popular' as a household and family ale. 'Other varieties were shown, especially the sparkling, wholesome, fine-flavoured mild ale, which we were told was in demand for a distinct section of the trade.'

The heady hoproom at the top of the building – 'what a beautiful aroma saluted our senses' – was occasionally used for grander purposes than storing the huge pockets of hops from Worcestershire and Kent. 'It was in this room that the firm entertained 200 of their customers and friends to luncheon on the day of the Queen's visit to Wrexham, which happy event took place a few weeks before our visit.'

Frederick Soames, like Peter Walker before him, had quickly climbed to the top of the social pile. In the year of Queen Victoria's and Barnard's visit he was Mayor of Wrexham (1891–2). He was also chairman of the Brewers' Association of North Wales and with his father and brother owned extensive ranges of maltings near Grimsby in Lincolnshire. In addition he was chairman of Edisbury's North Wales Mineral Water Company in Wrexham. Not a bad achievement for a man still in his early thirties.

Before leaving, Alfred Barnard was invited for lunch to Soames' stately home, Llwyn Onn Hall, outside the town. The author was even more impressed than he had been at the brewery:

> The house, which forms a prominent feature in the landscape, stands in a wooded park surrounded by a well-planted lawn and pretty flower gardens, and as we drove along through the domain, we caught many a glimpse through the tall trees of fields dotted with sheep and cattle and the shadowy blue of distant hills.

Frederick Soames knew how to live. He was later to become mayor twice again from 1901–3. Sometimes his stylish living seemed to take precedence over the demands of his business.

When called to give evidence before the Royal Commission looking into the working of the Welsh Sunday Closing Act in 1889, he argued that the Act had made drinking on the holy day worse. 'When the Act was passed I at once discovered that there was a great tendency to the increase of Sunday drinking in the various houses over which I have control.' (Travellers were allowed to drink on the Sabbath, so many men had taken to walking to distant pubs for the day.)

And what was Soames prime concern?

> It is this, that living near the town and having a garden, I require it to be looked after, and the policeman who is supposed to look after my district tells me that since the Sunday Closing Act he cannot come and look after my garden because he has to look after the public-houses on the Sunday.

The commission chairman passed on to the next point without comment.

A much more serious problem was the railway. On the south side of the brewery, the Wrexham and Ellesmere Railway Company wished to run its extended line right through the brewery stables. The railway company had powers of compulsory purchase under the Wrexham and Ellesmere Railway Act of 1885. The plan almost closed down the brewery.

In a court case in 1893 Frederick Soames' barrister argued that the loss of the land would make it impossible for Mr Soames to carry on his business efficiently and so the railway company must compensate him by buying the whole brewery. Only after much legal wrangling did the brewery survive. The line was built shearing the rear of the premises and restricting the room for expansion.

Soames became so dominant across North Wales – pubs as far away as Holyhead on Anglesey were advertising its beers after the turn of the century – that the brewery simply became known as 'The

The Ale that got the Medal

1927 BREWERS' EXHIBITION

(Bob Foster)

Welsh Ale Brewery'. A large tied estate was developed. When Frederick Soames took over the Wrexham brewery in 1879 it had only six pubs. By his death in 1926 it boasted over a hundred houses.

The company gained fame of an unusual kind during the First World War. One of their motorized drays, requisitioned by the army, was one of the first British vehicles into action in anger in France. It was a short-lived appearance – captured for posterity by a war photographer. The brewery wagon was hit by German shells during the Battle of the Marne, barely a month after the conflict broke out in August 1914. It had been rushed across the Channel to the front so fast to ferry ammunition that it still carried the brewery's name. The word Soames could be made out on the burnt-out lorry, abandoned in a desolate no-man's land of mud, tree stumps and barbed wire. Later the evocative photograph was used as a poster by the brewery under the title 'Soames at the Front, France 1914'.

After the war a new five-storey brewhouse was built fronting on to Mount Street. Completed in 1920 it was officially topped off by Julian Soames, one of the proprietor's sons. In the final stone was placed a list of the employees and details of the firm's brewing capacity. The crude brick building reflected post-war austerity, and many complained it was an eyesore, ruining the view of the graceful parish church.

After Frederick Soames died in 1926 the brewery seemed to lose direction, which was fatal in the depressed and difficult trading conditions of the 1920s and 1930s. Many other Wrexham breweries ceased trading at this time.

A trade directory of 1922 still listed eight breweries in the town including Chadwick's Burton Brewery, Heasman's Eagle Brewery and Sisson's Cambrian Brewery. But within a few months these three had gone. Sisson's closed in 1922, leasing its forty-eight pubs to the Island Green Brewery. The Eagle Brewery also flapped its final flight that year. Both were in Bridge Street where, ominously, Ind Coope of Burton-on-Trent opened a depot.

The third to vanish in 1922, Julius Chadwick's Burton Brewery, illustrates how the Wrexham brewers had lost faith in the post-war years. Hubert Chadwick had taken over the business from his father that year and a personal portrait in *The Country Illustrated* magazine revealed:

> The Burton Brewery at Wrexham has always been well known as the best place in Wrexham for beers of all kinds; and in view of the increasing difficulty of pleasing the public palate Mr Hubert Chadwick has decided to abandon brewing on his own account. He is convinced, after actual experience, that a certain well-known North Country brewer can brew beer even better than that formerly produced by the Burton Brewery; and he has accordingly made arrangements for the business to be converted into a bottling stores with retail offices attached.
>
> Having abandoned brewing Mr Chadwick has arranged with the firm of Messrs Walker and Homfrays Ltd, the well-known Manchester brewers, for supplies of their celebrated draught ales and stout, which is always good stuff, as Manchester's millions recognise, and is now taking well in the Wrexham district.

It was an astonishing admission. With friends like that, the local brewing industry needed no enemies.

Chadwick concentrated on bottling and the wholesale business. The Burton Brewery's pubs were sold to Soames, where changes were also underway. F.W. Soames was registered as a private company in April 1931, under another of Frederick's three sons, Major Evelyn Soames, in tandem with John Rankin, a partner in the firm of chartered accountants, Peat, Marwick, Mitchell. The firm had a nominal capital of £100,000.

The new company did not survive long. Two months later in June 1931 it amalgamated with its main Wrexham rival, the Island Green Brewery, and a Shropshire brewer, Dorsett Owen of Oswestry, to form Border Breweries. The new grouping was represented as a combination of strength. In reality it was a huddling together of troubled firms. Border was a child of the Depression.

DRAGON OF THE NORTH

Although F.W. Soames provided the larger brewery, the dominant figures in the creation of Border Breweries in 1931 seem to have come from the rival Wrexham company, the Island Green Brewery.

Founded in 1856 by John Jones on a low-lying site between Pentre Felin and the Central railway station, in an area known as Island Green (some say it was originally called Ireland Green), the brewery had developed a sound reputation for its mild and bitter ales. The company offered farmers a prize for the best local barley, which was then used in its own maltings.

The premises were off Brook Street in the appropriately-named Brewery Place, between Well Place and Watery Road. As the street names imply, this was a site well blessed with springs, and others had brewed there before John Jones. Thomas Evans is recorded as a brewer in Island Green in Pigot's Directory of 1822. The buildings, some of which still stand today, were quite extensive, the roofs boasting elaborate tiling from the continent. The ground was riddled with tunnels and cellars, around which sprang many legends.

Some said the stone slab tunnels led to the parish church, having been built by monks in the twelfth century as a refuge and escape hole. Others believed bricked-off passages led to the Shire Hall. More cynical observers commented that the winding ways were ideal for hiding barrels from the excise men.

A confident visitor to the 'labyrinthine cellars' in 1911 reported in the *Wrexham Advertiser*: 'The gas jets at intervals show deeper depths and corners which might contain awful secrets, if one did not know that they only contained casks of as purely brewed ale as it is possible to get.'

This was no backyard brewery but a sizeable concern, almost comparable with F.W. Soames' business a quarter-of-a-mile away. The introduction to Slater's Directory in 1895 mentioned them in the same breath: 'Wrexham is famous for the fine quality of its ales and there are several extensive breweries including John Jones and F.W. Soames.'

John Jones of Grove Lodge was a rich – and generous – man. Like other leading brewers, he developed a reputation as a benefactor of the town. He built St John's Church in Hightown at a cost of £8,500 and endowed it with a further £3,500.

In 1905 the business was bought by Francis Huntley and George Mowat. In 1925 they registered the company with twenty-three houses. Some pubs in the town still carry Island Green windows. They also leased forty-eight pubs, following the closure of Sisson's Cambrian Brewery in 1922.

Despite the depressed trading conditions of the period, the venture must have been reasonably successful since in 1930 the company was preparing plans for an extension to its fermenting room. Then it discovered another way to solve its brewing capacity problems.

Border Breweries was registered as a private company on 27 June 1931, combining the three breweries of Island Green, F.W. Soames and Dorsett Owen of Oswestry, to the south

CRUMBLING HISTORY: the ruins of the Island Green Brewery in the distance, with a boarded-up Border pub, the Old Three Tuns, in 1992

of Wrexham. The nominal capital was £440,000 (compared to £100,000 for Soames alone two months before).

An Island Green man took the top post, Francis Huntley becoming chairman, a position he held until 1944. George Mowat was a director along with Evelyn Soames. Captain Melville Cordeaux represented the Shropshire brewers. John Rankin, the other Soames director, was also on the board. He lived in Sheffield and really represented chartered accountants Peat, Marwick, Mitchell, who were appointed auditors of Border Breweries. At the time of the amalgamation F.W. Soames was in liquidation.

The early years were a struggle. The company had been named Border Breweries because its trading area straddled the border, reaching into Cheshire and Shropshire besides North Wales. In a bid to beat the Depression the company pushed further into England, heading east into the more populous Potteries using the odd slogan, 'The Wine of Wales'. Drinkers were told to look out for pubs 'where the blue sign swings'. The red dragon had yet to arrive.

The Island Green Brewery was soon closed, with brewing in Wrexham concentrated at the Soames site. After the takeover of the Colwyn Bay-based S.K. Williams chain of off-licences in 1936, the famous cool cellars at Island Green were used to store Border's wine and spirits department. The crates and cases were said to be visited by spirits of a different kind – a ghostly cloaked figure who was dubbed by workers 'John Jones'.

Brewing may have lingered longer at the Oswestry brewery in English Walk off Smithfield Road. Dorsett Owen's popular light mild won a silver medal at the London

IN THE SPRING-

A young man's fancy
lightly turns to thoughts
of—
Good Beer as a Tonic
and as a source of energy.

He naturally goes
*Where the blue sign
swings!*

BORDER ALES

The Wine of Wales

ROMANTIC APPROACH: a comical Border advert
from March 1936, when the beer was sold as 'The
Wine of Wales'. Two years later the Red Dragon
roared in with the claim to be 'The Prince of Ales'

Brewers Exhibition in 1938, causing the brew
to be renamed Exhibition.

As the economy recovered, Border went
public in 1938, selling shares to raise capital for
expansion. The rampant red dragon symbol was
introduced on a wide scale at this time to help
the company roar into life. 'Order a Border'
became the order of the day, along with the
title, 'The Prince of Ales'.

During the war Welsh drinkers were for a brief
period unknowingly introduced to an exotic brew
– smoked beer. In 1941 German bombers had set
alight the peat and heather on Rhos mountain.
Lord Haw Haw on German radio claimed their
planes had struck an oil depot, and in this belief the
enemy pounded the poor hill for two nights
'causing dense smoke for over a week that one
could hardly see across a street in Wrexham',
recalled one resident. Brewing had to be
suspended during this period, but brews in process
were contaminated by the smoke. However,
nothing could be wasted during the war, so the
tainted beer was blended with later brews and sold
to an unsuspecting public.

In 1951 Border bought a former clothing
factory in Holt Road, Wrexham, and turned it
into a bottling hall and store, after pressure for
space at the brewery in Mount Street became too great. The beer was transported the short
distance by tanker. Border's mineral water factory was later
transferred to an adjoining site from its premises in Salop
Road.

By now Evelyn Soames was chairman (1945–63)
with J.F.M. Mowat as managing director. The sales
director was Bill Kington, a grandson of Frederic
Soames. He was also father of the future newspaper
columnist Miles Kington. Only Mr Kington senior
did not envisage his seventeen-year-old son
becoming a famous journalist. He wanted him to
work at the brewery and packed him off to
Wuppertal in Germany to gain experience.

Young Miles stayed with a Mr Burgener, head of the
Wickuler Brewery, which his father had helped rebuild
after the war. The budding writer soon proved he already
had a wicked way with words:

ISLAND GREEN
BREWERY
CO.
Nº I
STRONG ALE
Brewed from
Malt & Hops
ONLY
WREXHAM

One day Mr Burgener took me off to the Wickuler Brauerei, which was an enormous place compared with Border Breweries in Wrexham. He was terribly proud of it, as he might well be since he had built it up in a dozen years since the war. The vats glittered and the cellars stretched gleaming for miles, not like the friendly but rather ramshackle buildings of Border Breweries, which might have benefited from being devastated by the German army and rebuilt.

'Is your father's brewery so big?' demanded Herr Burgener.

'Not half so big,' I admitted.

'Does he have so many cellars?'

'Not so many.'

Mr Burgener's satisfaction at having the bigger brewery soon started to turn to complacency, whereas I began to feel indignant on behalf of my father. Had my father not come over specially at the end of the war to get this brewery started? Was it right now for Mr Burgener to have a bigger and better brewery? I did not think so.

So when he started quoting production figures at me, I started doubling them and passing them off as Border's. Did Wickuler produce 10,000 barrels a week? Then Border did 20,000, and so on. Mr Burgener's brow grew quite furrowed with the attempt to understand how a Welsh brewery half the size of his could effortlessly produce twice as much beer.

The harsh reality was that the company faced an uphill task in a difficult trading area, even though Border was now the only ale brewery in Wrexham since Beirne's Albion Brewery had ceased brewing in 1937. When a reporter from the *Wrexham Leader* newspaper visited the brewery in 1950, he found the managing director, Mr J.F.M. Mowat, in a downbeat mood.

There was 'an appreciable falling off' in demand for beer, said Mr Mowat, though this experience was nationwide. 'We are satisfied that the public cannot afford to pay for it. We welcomed the fact that we could make stronger beer [after the ending of wartime restrictions] but we are not able to produce cheaper beer. Eight pence out of every shilling paid for a pint goes to the Chancellor of the Exchequer.'

By the 1960s Border owned more than two hundred pubs and seventeen off-licences, but they were scattered over a wide area and many houses did little business, particularly in winter. Much of the trade was seasonal, with heavy demand only in the busy but all-too-brief summer holiday months. As a later head brewer said about his houses: 'Some are very long hauls, half way up a mountainside. We have a very far-flung empire.'

Could they stand alone in the face of stiff competition from Britain's national brewers? Even the Wrexham Lager Brewery, which had bought Beirne's twenty-five pubs in 1938, had sold out to Ind Coope of Burton-on-Trent in 1949. In 1961 Ind Coope had become part of the huge Allied Breweries combine. Takeovers were tearing through the industry.

In 1962 Border decided to shelter from the storm under the Whitbread umbrella. A statement just before Christmas declared:

The directors of Border Breweries are pleased to announce that they have concluded a long-term trading agreement with Whitbread and Company Ltd, covering technical and commercial relations between the two companies. Whitbreads have acquired a substantial interest in the equity of Border Breweries, whose directors are confident that this association will be of great advantage both to the shareholders and to those who work for the company.

Border by now employed some 250 people.

What had finally forced Border's hand was that it had issued easily-traded, five-shilling shares in the summer. Placed at 11s. 9d., they had shot up in value to 19s. Someone was buying heavily and could soon pounce to take control. Border quickly sought sanctuary with Whitbread, which the previous year had merged with Flowers Breweries to form a giant £73 million group.

The tie-up gave Whitbread a 19 per cent stake in Border and a seat on the board. Its beers like Tankard also went on sale in Border pubs. Inevitably within a few years there were rumours of a full takeover, particularly since Whitbread was snapping up breweries right, left and centre including two in South Wales, Rhymney Breweries in 1966 and Evan Evans Bevan of Neath in 1967.

Yet Whitbread did not move in for the kill in Wrexham. Border may have been saved at this stage by its own peculiar problems. Its many low-turnover pubs were not that attractive in economic terms, especially since Whitbread by the end of the decade was choking on the vast estate of 9,000 houses it had acquired through takeover.

Even less of a tempting morsel was the increasingly cramped brewery in Mount Street. Frederic Soames' problems with the railway company were coming home to roost. The track at the back meant there was no room for expansion. When Border tried to extend upwards in 1963, they were blocked by the Town Council. Brewers no longer held sway on the local authority. The councillors ruled that the brewery would not be allowed to interfere any further with the view of the parish church.

Whitbread was certainly not interested in the premises. Chairman Bill Whitbread revealed in his annual statement in 1971 that the group in the previous ten years had closed 15 breweries, 24 bottling plants and 54 distribution depots. It had more than enough brewing capacity, especially since it had retained and enlarged ten of the breweries it had acquired, including Threlfalls in nearby Liverpool.

There was another reason for Border's survival. There were still a few breaths of life in the red dragon. Whitbread claimed it only took over breweries 'by invitation'. The Border board was not yet ready to send out invitations to its own funeral.

Mr C.F. Huntley was chairman from 1963 to 1978, and under him the wholly-owned subsidiary of S.K. Williams of Colwyn Bay diversified into supplying a wide range of other brewers' beers from outside the area, acting as wholesalers for the free trade along the North Wales coast as far as Anglesey. Depots were opened at Mochdre, Llanllechid and Towyn. 'The coastal holiday area attracts people from many regions of England and Scotland,' said S.K. Williams' sales director Raymond Formstone, 'and they like to be able to buy beer of their choice'. Fears began to grow, given Border's problems at its brewery, that the company might pull out of brewing altogether and just concentrate on selling other brewers' beers.

Border seemed reluctant to push its own brands. It produced a Border Bitter and two milds – a dark 3X and the light Exhibition. All were low-gravity, relatively weak brews, characteristic of a region once dominated by collieries where the prime aim was to slake the thirst of the miners. None had a high profile or much promotion.

The arrival of CAMRA, the Campaign for Real Ale, in the early 1970s had given a boost to independent breweries selling traditional cask-conditioned beers, like Brain's of Cardiff. But Border failed to benefit from this revival of interest since it had been moving sharply the other way.

Border's policy was to introduce processed keg beer into its houses. An internal CAMRA report of 1978 revealed: 'Three years ago 83 of their pubs still served traditional draught ale. Today only 37 houses serve it, and some of these have only bitter or mild.' It was not just the low-turnover country pubs that were being switched over. Only two Border houses in Wrexham served real ale. The report concluded that the brewery 'continue to swim against the real ale tide'.

SLEEPING DRAGON: only late in the day did Border wake up to the market for real ale

Such was CAMRA's concern, the consumer group arranged a meeting between its national executive and the brewery's directors in Wrexham in 1979. Over a pint in the Nag's Head the two sides discussed their differences. The company strongly denied that it was planning to stop brewing and blamed their move away from traditional beer on the club trade.

The area has a large number of clubs, partly due to the Sunday Closing Act that for many years shut pubs in Wales on Sunday. Head brewer Don McIntyre revealed that 'a large proportion' of their production went to the clubs. 'We could not exist without it.'

The club committees and stewards wanted convenient, easy-to-handle beers. So in the late 1960s Border began to filter its beers for the cellar tanks that the larger clubs installed. Then in the early 1970s it began to pasteurize its draught beers as well, so they could be available in smaller containers – kegs – for smaller clubs and pubs.

In 1979 it even introduced its own lager, Keltic Lite – the maturing tanks being introduced at the expense of part of its cask storing and filling area. 'There's nowhere we can go', said Mr McIntyre. 'We are already over utilising space and can't go upwards because of the church.'

Just as worrying for the CAMRA party was its tour of the brewery. All seasoned brewery visitors, they were surprised by the poor state of the plant and the buildings. The impression was confirmed when the second brewer Alan Beresford took some aside and admitted: 'If I was 20 years younger I would be very worried indeed. These coppers are wafer thin.' One, moved from the Island Green Brewery, was over eighty years old.

In the basement the nineteenth-century vaults were crumbling under the strain of passing heavy lorries. The writing was clearly on the brewhouse wall. Without large-scale investment the brewery would eventually close.

A feature in CAMRA's newspaper *What's Brewing*, under the headline 'Bordering on the edge of oblivion?', concluded:

FLYING THE FLAG: Border Brewery in the early 1980s with Soames' proud 1894 chimney between the Nag's Head on the left and the crude 1920 brewhouse. On Marstons' takeover in 1984 the towering chimney was found to be only a quarter-of-an-inch out of line after ninety years

The big worry now is not so much that there is a continued drive towards keg (that seems to have been blunted although not reversed), but rather that there might not be enough drive within the company itself to ensure the continued existence of the last remaining independent brewery in North Wales.

Despite the prophecies of impending doom, Border celebrated its fiftieth anniversary in 1981 with a special bottled Golden Jubilee Pale Ale. A twelve-page supplement in the *Liverpool Daily Post* boasted that 'Border has grown until today it has a turnover exceeding £14 million and trading links with Whitbread, Allied Breweries, Bass and other concerns.' New managing director Joseph Hatton added: 'We are confident that we can build on the foundations laid in the past and secure a successful future.' It was to prove a short-lived future.

On the beer front, there were a few pint-size developments. In 1982 the company made a belated move to capture some of the growing real ale market with the launch of a premium bitter called Old Master. The dark XXX mild won a gold medal at the International Brewers Exhibition in 1983.

Border also boasted an unusual claim to fame, unmatched by any other brewery – they owned the local football club ground of Wrexham FC, called the Racecourse, which perhaps explains why it is the only major pitch in Britain with a full-blown pub kicking about in the terraces. The busy two-bar house, The Turf in Mold Road, still serves the faithful followers today.

But overall trade was difficult, with continuing high unemployment in the area. At times the brewery was on a four-day week, or less. Stockbrokers Grenfell and Colegrave rated the company a poor investment.

Then the shares took a sudden leap upwards, shooting up 43p to 155p in a day on 15 February 1984, following rumours that a bid had been made for the brewery. The shares rose even higher to 208p when it was revealed that there were two rival bidders.

Burton brewers Marstons, best known for their Pedigree bitter, had made the original approach. Burtonwood Brewery of Warrington, who had been courting Border for months, had then stepped in with a £9.4 million cash offer. 'Their approach flushed us out', said Burtonwood's chairman Graeme Dutton-Forshaw. 'When it was announced, we decided it was time to move because we did not want to let Border slip through our fingers.' Both breweries already owned houses in North Wales, Burtonwood having ninety in the region.

The Warrington company's offer contained a significant pledge about the Wrexham brewery: 'All viable production operations will continue at Border.' Marstons made no such promises.

Whitbread was in a crucial and central position in the battle for control. The brewing giant owned 19 per cent of Border – and 35 per cent of Marstons. Its influence was decisive. On 8 March Marstons clinched a £13.8 million takeover, sparking an angry reaction from Graeme Dutton-Forshaw. The Burtonwood chairman talked about deals 'put together behind closed doors' making it impossible for him to compete. He complained to the Office of Fair Trading, but the takeover stood. Besides Whitbread, Marstons also obtained early acceptances from the Border directors, who owned 14 per cent of the shares.

In October the inevitable announcement was made. The Wrexham brewery was to close, with the loss of around 120 jobs. The beers for the 170 Border pubs would be supplied from Marstons' brewery in Burton, where some 'Border' ales would be produced. The bottling plant building in Holt Road was retained as Marstons' local offices. 'It is an act of bad faith', said Wrexham's chief executive Sydney Tongue. 'It is extremely upsetting that this local institution with 250 years of brewing history behind it is to close.' Two years later in 1986 the closure still rankled with Miles Kington, who had turned down the chance to work there:

And if I had decided to go into the brewery after my father, I would now be without a job. Two years ago Border Breweries were taken over by Marstons and the old friendly brewery, with the railway line running on a bridge across it and the red Ruabon brick chimney towering above it, was closed down.

This Christmas it was set on fire, probably by homeless squatters, and is now a gutted derelict site, no longer steaming, smelling of yeast day and night, and resounding to the crash of rolling barrels. It looks just as the Wickuler Brauerei in Wuppertal must have looked in 1945, in fact, but this time I'm glad my father wasn't around to see it.

LAGER LEADER

Today lager accounts for more than half of all the beer brewed in Britain. Yet in 1960 only a few brewers bothered with the 'foreign froth', and a hundred years ago hardly any British drinkers had heard of it, let alone drunk a drop.

The first determined attempt to convert these dark islands of ale on to the golden path of lager was made in Wrexham and London. It was not to prove an easy experience for either of the ventures.

The tortuous Welsh tale began in Manchester around 1880. A group of wealthy men, led by German and Czechoslovakian immigrants, shared a common prejudice and a passion. They disliked warm English ale and yearned for the cool lager of their homelands. Once a fine pilsener was produced, they were convinced native drinkers would flock to their brewhouse door. Had they foreseen the heavy hangover ahead, they probably would never have bothered.

To fulfil their dream it is significant that they came to Wrexham. It is a mark of the town's reputation for brewing excellence. When they registered their new company on 6 May 1881, with offices in Manchester, they were proud to name it the Wrexham Lager Beer Company.

Among the seven directors were manufacturing chemists Ivan Levinstein and Otto Isler and a banker Maurice Schlesinger, all from Manchester, plus two English textile merchants. The new company was limited to £50,000 in 5,000 shares.

On 28 April 1881 another chemist director, David Johnson of Chester, had already bought on behalf of the venture a two-acre plot of land off Holt Road, to the east of the town centre, for £3,500. The site must have looked sound as it was near the famous Penadur Springs, which supplied Edisbury's mineral waterworks. However, it was soon found that while the water was suitable for ale brewing, it was no good for lager.

Tests were carried out to find the right brewing water around the town. They discovered that one area, Mill Pond Meadow to the west, offered similar water to Pilsen, the Czech town which had pioneered pale lager brewing forty years before.

The company decided to build its own brewery by this source on St Mark's Road (later Central Road), slightly set apart from the then town centre, but close to both main railway stations. The ground cost £3,750 and provided soft spring water and a hillside in which to cut the deep underground cellars in which to mature the beer.

The original site may not have been a dead loss entirely, as alongside were the buildings of the Victoria Brewery, which had gone into compulsory liquidation in 1878. The equipment, however, was still in place, and some believe this was used for trial brews of lager. The premises later became Allmand's Victoria Flour Mills.

Engineers from Austria drew up the plans, based on a brewery they had already built on the continent. Work on the buildings began in March 1882 at a cost of £20,926. A further £10,748 was spent on plant and machinery. This was no ordinary operation, in a town already bursting with breweries, as the *Wrexham Advertiser* newspaper reported on 20 October 1883:

> The progress of this extensive undertaking has been watched with considerable interest by the public generally for some time past, and now that the great bulk of machinery is in position, it was deemed a fitting opportunity for the shareholders to pay a visit of inspection to the works.
>
> A general assembly of the directors and shareholders of the Wrexham Lager Beer Company was accordingly held at the Wynnstay Arms Hotel on Saturday (October 13), under the auspices of the President of the Company, Mr Ivan Levinstein; the party first being conducted through the new works by the general manager Mr Stanislav Fenzl. Flags in honour of the occasion floated from the summit of the brewery.

Mr Fenzl was responsible for the plans, which included features never before seen in a British brewery. In particular '200 big lager casks (each holding 1,446 gallons) are laid in six lager cellars, each 60 feet long, 10 feet wide and 19 feet high'. These cellars were to be cooled to near freezing point by ice produced by an ice machine, capable of turning out 5,000 tons a year.

Other special points were continental decoction mashing, the use of bottom-fermenting lager yeast and double fermentation, first in thirty fermenting tuns each holding twenty-five barrels and then into the ice-cold lager cellars for 'a few months' for a long, slow, secondary fermentation to produce a clear beer that 'possesses great durability'.

The *Wrexham Advertiser* article explained the thinking behind the pioneering project:

> It is desirable to check the continuously increasing importation of lager beer to England and the Colonies from the Continent and America by home production, and thus re-obtain the markets for English beer where the heavy-bodied English ale has been displaced by foreign lager as in Turkey, Egypt etc.
>
> About 40 years ago lager beer brewing was cultivated only in one German province. Since that period, however, the whole of Germany, Austria, Scandinavia, Holland, Belgium, France, Russia, Italy, America and even Japan have adopted lager beer brewing. The greatest obstacle in England was the difficulty in procuring cheap ice. This difficulty is now put aside by the excellent ice machines manufactured today.

Added the reporter: 'We understand that brewing operations will commence in a month or two.' By the end of 1883 the Wrexham Lager Beer Company was buying large stocks of barley (the brewery had its own mechanical maltings) and hops. Some £1,963 was spent on beer's raw materials. Brewing had begun – and so had the headaches.

The company was heavily mortgaged and the period between the first investment and the start of trading was proving a major strain. Worse, its golden dreams were dashed. On the continent, the lager cellars were kept freezing cold by packing them with thick ice cut from the lakes. In the milder Wrexham winters, the much-vaunted ice machines struggled to

UNCERTAIN OCCUPATION: for the early brewery workers, the troubled first decade meant their jobs were never secure (Wrexham Lager)

produce the same consistently chilling effect. Their failure meant that the company could only produce dark Bavarian lager, not the pale variety it had intended. Worse still, it had underestimated the local drinkers' attachment to their celebrated ales. Wrexham's packed pubs continued to enjoy their milds, stouts and pale ales, and largely ignored the strange new brew in town.

It was during this difficult period that one of the leading directors, Ivan Levinstein, bumped into the formidable figure of Robert Graesser while travelling by train between Liverpool and Manchester in 1886. Another manufacturing chemist, Graesser had founded the chemical works at Cefn Mawr near Wrexham, later to become Monsanto Chemicals.

An inspired innovator – he held the patent for the distillation of tar, the basis of asprin and camphor – and a driving businessman, he was just what the ailing Wrexham Lager Company needed. He was invited to join the board and the company was re-formed with Robert Graesser putting in £7,260, making a total subscription of £30,000.

A private note circulated in September shows that the company already had 1,300 customers on the books, but they were scattered and mainly small beer, many taking bottles only. Sales for the three months to 31 August 1886, were 20,550 gallons – less than 50 barrels a week. Despite this limited business, the brewery did enjoy the distinction of taking the silver medal at the Liverpool Exhibition that year.

Even though the brewery was a sideline, Robert Graesser was determined to make it a success. First he tackled the brewing difficulties by introducing a major technical advance for the time – mechanical refrigeration. The process was then in its infancy, but he was familiar with it as he had already used refrigeration at his chemical works to produce phenol.

The installation of brine coils lowered the temperature in the cellars, improving the quality of the beer and enabling a full range of lagers to be produced. The only drawback was when the system needed defrosting – huge lumps of ice would crash from the ceiling.

Sales of the dark, light and pilsener lagers were still limited however. Most of the pubs in Wrexham were by now owned by the competing ale breweries, and Wrexham Lager had to content itself with distant business through private hotels and restaurants.

How far the company went in search of outlets is shown by a trade advert of 1888, which boasted of agents in Glasgow, Edinburgh and Dublin, besides stores in Manchester. It also illustrates an interesting marketing ploy. Lager was presented as a temperance drink! This had been a theme even before the company had started brewing. An early presentation claimed that 'as a golden mean between strong English ales and strictly teetotal beverages, lager beer will no doubt be welcomed both for its nourishing and refreshing properties'. The 1888 advert developed this line, claiming that Wrexham Lager 'contains the maximum of extractive malt liquor with the minimum of alcohol, combined with all the principal tonic

THE WREXHAM LAGER BEER CO., Ltd.

BREWERY :—WREXHAM, NORTH WALES.

Manchester Office and Stores--14, Brown St.

BREWERS and Bottlers of Lager Beer of the finest quality, surpassing any other Lager imported or offered in this country. It contains the maximum of extractive Malt Liquor, with the minimum of Alcohol, combined with all the principal Tonic qualities of the finest Hops, being brewed from the finest Malt and Hops. it is a splendid tonic, and a light, wholesome, and refreshing drink.

It gained the only Medal awarded to Lager Beer at the Liverpool Exhibition, 1886, and was the only Lager Beer sold at the Royal Jubilee Exhibition, Manchester, 1887.

If you cannot obtain it in your district, send direct to the Brewery.

Agent for Glasgow and West of Scotland—

A. McGUFFIE, 79, West Regent, Glasgow.

Edinburgh and Leith—

JOHN USHER MENZIES, 10, Bernard St., Leith.

Dublin and District—

W. M. WORRALL, 27, Percy Place, Dublin.

HARD SELL: an advert of 1888 showing that Wrexham Lager was already pushing its 'light, wholesome and refreshing drink' as far as Dublin and Edinburgh

qualities of the finest hops. It is a splendid tonic, and a light, wholesome and refreshing drink.' Wrexham Lager's early slogan was 'absolutely pure and wholesome'.

The lager even received the official blessing of the temperance societies, as recorded in a Certificate of Purity:

> The Wrexham Lager Beer Company has been successful in producing a light Pilsener Lager Beer which not only refreshes but acts as a tonic in cases of weak digestion, and is almost non-intoxicating. When more generally known and consumed, it will diminish intoxication and do more for the temperance cause than all the efforts of the total abstainers.

There was some small grain of truth in this. Professor Charles Graham of University College, London, in a paper on lager presented to the Society of Chemical Industry in 1881, compared British ale and continental lager in detail. He found that while British ales of the time had a heavy original gravity of around 1065 (nearly 6 per cent alcohol by weight), lagers had a lighter gravity of around 1050 (4½ per cent alcohol).

This would be regarded as a premium strength beer today, but a century ago it was almost viewed as non-alcoholic. Indeed, one South Wales advertiser took just this attitude in 1891. Wine merchant and beer bottler, W.F. Bull of Swansea, the sole agent in the area for Wrexham Lager, sold it as 'the prize medal non-intoxicant'. Some were convinced.

Lager was becoming much more socially acceptable than ale among the more refined classes. It was sold at the Royal Jubilee Exhibition in Manchester in 1887 and at the National Eisteddfod in Wrexham in 1888. The champagne-shaped bottles graced the top table at Bromsgrove School in the 1890s. There was only one trouble with this delicate market. It didn't drink enough. The Wrexham Lager Beer Company went into liquidation in 1892.

The company had not been helped by a rival. Despite Wrexham Lager's later claim to be the first British lager brewer, it had been narrowly beaten to the punch by the Austro-Bavarian Brewery and Crystal Ice Factory of Tottenham in London.

As its name implies, this venture had also been launched by German immigrants under one Leopold Seckendorf. Formed in March 1881 it was brewing by late 1882 on a large eight-acre site off Tottenham High Road, which doubled as a cold store. Wrexham Lager later claimed 1882 as its anniversary, but it did not brew for sale until late 1883.

From its earliest days Austro-Bavarian pushed the 'health drink' angle, winning the highest award for its beer at the International Health Exhibition in 1884. But, like Wrexham, the London lager pioneers found trade difficult, and began to search wider and wider for a viable market.

In 1886 the company was reformed as the Tottenham Lager Beer Brewery and Ice Factory, gaining further honours around the globe – a First Order of Merit in Adelaide, Australia, in 1887, and a Gold Medal in Paris for its Rainbow Lager in 1889. Numerous brand names were registered from Angel and Mermaid to Dice, Foot and Snail Lager! Tottenham even pushed into the heart of Wrexham Lager's territory, appointing E. Manners of the Carnarvon Castle Stores, Wrexham, as sole agent for North Wales.

All this frantic activity proved of no avail. The receivers were called in during the summer of 1894, a shareholder, Mr Wallraf, claiming that the company had gone on trading when it knew it was 'absolutely insolvent'. He said he was owed £30,000, a huge sum in those days. Tottenham eventually tottered in 1895, going into voluntary liquidation. A new company, Imperial Lager, was formed early the following year, but only lasted until 1903 when brewing ceased.

Wrexham Lager could just as easily have stumbled and collapsed as the Tottenham company – except for the perseverance of Robert Ferdinand Graesser.

Born in Mosel, Saxony, in 1844, he had studied chemistry at Chemnitz before coming to Britain in 1863, at the age of nineteen, to work in a Manchester laboratory. Four years later he was setting up his own chemical works in North Wales at Cefn Mawr near Ruabon.

His first partnership dissolved after a quarrel and he lost his first business in extracting paraffin from coal shale when new American oil wells flooded the market. So he turned to tar distillation from coal and then to the production of pure phenol from crude carbolic. By 1887 he had virtually rebuilt the plant three times and boosted output from 20 to 1,000 tons a year. By 1910 the firm was responsible for half the world's supply of phenol.

Later Monsanto Chemicals of St Louis, Missouri, who took a half share in Graessers in 1920, paid this tribute to him:

FORMIDABLE INDUSTRIALIST: Robert Graesser made Wrexham Lager work

> Many men faced with the difficulties that beset Graesser during his early period would have given up, or at least have sought help. But he was a man of immense character, determination and faith in himself.

Robert Graesser was never one to quit. When, at an extraordinary general meeting on 30 September 1892, it was agreed to wind up the Wrexham Lager Beer Company, he resolved to continue. He bought the assets from the liquidator on 11 October, cleared the outstanding debts, and carried on trading.

Since the home market was sparse, Graesser decided to follow the flag into the British Empire. The improved stability of the beer meant it could keep for long periods, even in tropical heat. Graesser aimed to make it the constant companion of colonial officials and their thirsty troops.

The many British trading companies of the period saw the advantage of having their own brands, and Wrexham Lager duly obliged by selling its brews under a variety of exotic labels like Drummer brand for Richardson of London, Crocodile brand for Jaffe & Sons of Manchester and Howard's Five Fleets lager, popularly known as 'FF' brand. A special bottling store had to be opened in Warwick Street, Liverpool, to cope with this growing export trade.

The army was also impressed. Wrexham was a major garrison town, and soon the bottled lager was advancing with the troops around the globe. The brewery's own Ace of Clubs trademark found itself dealt into some desperate battles. A letter dated 21 September 1898, from a Sergeant Major serving in the Sudan, was received at the brewery in 1899. It read:

Gentlemen, I enclose herewith one of your labels which was taken off a bottle found in the grounds of Gordon's Palace at Khartoum on the 3rd September, 1898. I send it as a matter of curiosity, just to let you know how far your famous Wrexham Lager Beer can be had.

PUSHING THE PILSENER: by 1905 Wrexham was claiming to be 'the oldest lager beer brewers in the United Kingdom'. When this label was registered in 1898 Wrexham was in North Wales

It was not only the beer that played its part in this historic siege. The chaff from the crates in which the bottles were packed was said to have been fed to General Gordon's horses to keep them alive.

By the turn of the century 80 per cent of Wrexham Lager's output was for export. The lager flowed around the globe including Australia, Bermuda, Brazil, Ceylon, China, the Cameroons, Dahomey, India, the Ivory Coast, Java, Japan, New Zealand, Nigeria, Sierra Leone, Togoland, many South American states and the Pacific Islands. Most was sold under appropriate traders' labels like Kangaroo Lager, bottled for Hibberts of London, for the Australian market.

In 1900 Wrexham Lager was re-formed as a limited company, with capital of £50,000. Robert Graesser was still firmly in control. His eldest son and daughter were the shareholders. Six years later his other four sons also took a stake in the company, two managing the Liverpool bottling stores. Wrexham Lager was now very much a family affair.

Robert Graesser discovered another useful market when, in 1904, he made his first trip to America on the White Star liner SS *Baltic*, and insisted on taking a supply of draught dark lager with him. The beer survived the rigours of the voyage so well that the shipping line decided to stock it. Others followed and a lucrative business of supplying ships' stores with draught lager developed, especially for the large cruise liners of the day.

In 1911 Robert Graesser died at his desk, having secured Wrexham Lager's future – and its place in history. It was one of the first purpose-built lager breweries in Britain and the first to install refrigeration. It also boasted a special export trade, much desired by others.

Its export beers were the golden Wrexham Pilsener and the dark, Munich-style lager, which was the first beer brewed by the company. At home it also supplied a light lager and an unfiltered dark.

One of the great breweries of the day, Barclay Perkins of London, which was considering brewing lager itself, bid for the company shortly before the First World War. It wasn't just the plant and business it was after, as much as the expertise.

Ernest Graesser was the new governing director after the death of his father, but it was his younger brother, Edgar, who ran the brewery. He had obtained a degree in brewing in Germany and had spent several years working in a German brewery. 'There was certainly nobody in this country who had his knowledge', said one observer. Barclay Perkins' offer was conditional on Edgar Graesser remaining as brewing director. The family refused and after the war Barclay went ahead in 1921 and brewed its own London Lager, which proved to be a keen competitor overseas.

Wrexham's other main rivals were all Scottish, apart from Allsopps of Burton-on-Trent. Tennents of Glasgow brought in two brewers, a German and a Dane, to begin lager brewing in 1885, and by 1891 had built a completely new lager brewery alongside its existing plant. Jeffreys of Edinburgh followed in 1902, using Tennents' German brewer, Jacob Klinger. Arrol's Alloa Brewery began brewing lager in the 1920s after being taken over by Allsopps in 1921, following the burning down of Allsopps' own lager brewery. All these companies became major players in the export market.

On the home front, Wrexham Lager was beginning to pick up a local following. Miners coming home from the pit developed a taste for the dark lager, rich in malt extract. They would call in at the brewery for a drop and also pay a penny a pint more for a drink most today would send straight back to the bar – cloudy unfiltered lager from the bottom of the casks, specially sold in swing-top bottles, believing that with its high yeast content they were receiving a meal as well as a drink.

In the First World War there was a real danger that anti-German feeling could damage Wrexham Lager's slowly growing local trade. Their brewer, Justus Kolb, was interned on the Isle of Man, along with engineer Wiederman. Edgar Graesser took over the brewing duties. His nephew Bill Graesser-Thomas later recalled: 'Even something looking like a German sausage in a window was enough to have the shop wrecked.'

The company did themselves a power of good when, in 1917, there was a big fire at Thomas's Timber Yard, near the railway station, which was threatening to spread to the adjoining munitions work of Powell Brothers and Whittaker. The brewery dray-horses were called out to haul away stacks of wood from the blaze.

Leading local brewers Soames tried to cash in on the anti-German feeling in the army town, by producing a new beer called Sobright, describing it as lager-like – but not a German product. The Graessers responded by proclaiming: 'Wrexham Lager, British brewed since 1882. Refuse any imitation.'

Wrexham Lager's relationship with the local ale brewers was often strained. Some allowed their pubs to stock the lager, since it was not a beer they produced themselves. Others regarded the foreign brew with suspicion. Landlords acting without authority and taking the lager caused friction. In 1920, not for the first time, the local brewers banded together and boycotted Wrexham Lager.

With such barriers in North Wales, it was easier to find Wrexham Lager in London, in top venues like the Carlton and Constitutional clubs and the Great Western Hotel, than it was to drink a drop in Wrexham.

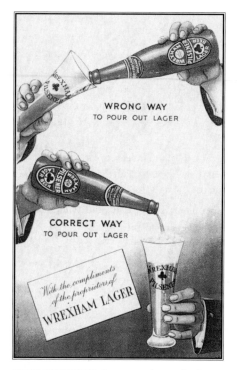

WRONG WAY
TO POUR OUT LAGER

CORRECT WAY
TO POUR OUT LAGER

FRESH EXPERIENCE: lager was such a novelty that Wrexham Lager between the wars issued leaflets to its customers telling them how to pour the beer

Some drinkers went to enormous lengths to satisfy their thirst. Crates of the dark lager used to be stacked in a corner of the bottle-room just below a hole in the wall. It was regularly noted that some of the top bottles were missing. Eventually, two men were caught fishing through the hole with a noose on the end of a line, trying to catch the swing-top bottles.

Trade continued with Africa and Asia and other distant locations and the brick brewery buildings, with their added fairy-tale Bavarian castle turret, retained an exotic flavour. Norman Graesser, grandson of the founder, particularly remembers their Egyptian agent.

He used to have very long little finger nails. We were as children told that this was to prove he never did any manual work. We used to snigger about this reason, as we thought the real purpose was to clean out his ears.

Wrexham residents became used to witnessing other strange sights, like giant casks, higher than a man, rolling along the road. These were the oak storage vessels, which once a year were cleaned and sprayed inside with hot resin, and then turned to give an even sterile coating by trundling them along the road outside the brewery, until the resin cooled.

In 1922 Wrexham Lager made an important breakthrough into the local trade by buying its first house, the Cross Foxes in Abbot Street, near the brewery. It may only have been one pub, but it was soon selling a remarkable thirty barrels of lager a week.

The beer was served through pipes that ran through lead-lined boxes packed with ice from the brewery's refrigeration plant, keeping the beer cool even on the hottest day. Wrexham Lager had scored another first – the first to serve genuinely chilled draught lager.

The manager of the Cross Foxes was the son of Sam Johnstone, the owner of the Bridge End Hotel, Llangollen. Bill Graesser-Thomas recalled:

He was a Wrexham friend if ever there was one. At that time Llangollen was a popular excursion spot from as far afield as Liverpool, Manchester and Birmingham. Coachloads of people were delivered to the town daily throughout the summer. The Bridge End was a very popular rendezvous. They had about a dozen bedrooms and the quantity of lager sold there was fantastic. I think Sam Johnstone at one time had four pumps going steadily and it was just about the only beer he would serve.

With flagship outlets like these, sales at last began to grow in Wales. But the main business of the brewery remained its foreign trade, despite a set-back in 1920 when Lever Brothers took over one of its main trading partners, the United African Company, which had already absorbed four British firms.

Lever Brothers, in the growing temperance mood of the day, discouraged the sale of alcohol. Fortunately, this loss was made good by the Elder-Dempster Company shipping out Wrexham's draught dark lager from Liverpool, for transhipping to its coastal steamers, besides handling its Ace of Clubs bottled pilsener.

Different brands were developed for different countries. In the new Irish Free State between the wars, feeling ran high against the British and British beer. So Wrexham introduced Graesser's Club Lager, diplomatically hiding its country of origin while proudly stating on its label, 'Bottled in the Irish Free State'. In America, Four Aces lager did the trick. Labels sometimes proclaimed 'brewed in Wrexham, England', to add to the confusion.

The siting of the brewery alongside the railway line, with its own siding, meant wagon loads of barrels could easily be transported to the bottling stores at Liverpool. Two trucks were regularly hauled out of the brewery – by shire-horses – and then shunted on to the Paddington express passenger train called *The Zulu* at Wrexham Station bound for Birkenhead.

When they were stocking up giant liners like the *Mauretania*, off to sea for three months, whole trains were dispatched overnight carrying 1,800 small casks in twelve covered wagons

STEAMING SUCCESS: a train picks up wagon loads of Wrexham Lager for the passenger liner trade in the early 1930s
(Wrexham Maelor Museum)

packed with ice. Business became so brisk, recalled Norman Graesser, that one Christmas period in the 1930s three complete train loads of barrels were dispatched to London and Southampton to supply cruise ships. He especially remembers one batch.

One morning in the late Thirties there was such a commotion outside and inside the brewery, with Customs & Excise men everywhere. Evidently 350 gallons of dark lager, after a 12-month cruise, had been returned to Liverpool to be put into trade. The gravity did not tally with that which was exported.

It was agreed in the end that the head brewer, sales director and myself should go over to Liverpool, together with the agent, to try and solve the problem. We suspected what had happened which proved correct. The water in the beer had separated during the long period of refrigeration and formed pieces of ice. They had taken a reading of the concentrated liquid.

We thawed out one four-and-a-half gallon cask and the mystery was solved. Needless to say, we had a party on the dockside, with all contented with the result.

The remainder of the 350 gallons were enjoyed by staff for their rations, and by the few fortunate customers lucky enough to buy some. 'They had a real vintage drink.' The beer was over eighteen months old.

Under Edgar Graesser, a brewing perfectionist rather than a businessman, the strict standards of the German pure beer law, the Reinheitsgebot, were maintained, as was a high gravity of 1052 for the pilsener and the dark lagers at a time when British beers had declined substantially in strength following wartime restrictions.

Local barley malted on the brewery's own floors was used, along with hops from Europe (and any continental varieties that could be obtained from the English hop experimental station). The yeast strain came from Pilsen.

This emphasis on quality proved a handicap in the depression as customers began to look for savings. The White Star line was offered Barclay's lager at 1s. 10d. a gallon, 2d. a gallon cheaper than Wrexham's price. At a meeting in London in 1929 Edgar Graesser told White Star's buyer, Mr Palmer, that it was unthinkable to compare Barclay's with Wrexham and bluntly refused to meet the price.

Some of the White Star business was subsequently lost to Barclay's. When Cunard made a similar approach, a wiser Edgar Graesser fell into line and held on to the trade. He even introduced a Club Lager at 1045 gravity for those wanting a cheaper drink, but he was never proud of the brew.

On one memorable visit, later recalled by Bill Graesser-Thomas, the pair of them called on the managing director of London brewer Meux, who was taking Club Lager in its houses. Edgar Graesser said he would prefer to supply Wrexham Pilsener at the same price rather than Club.

The reply of Meux's managing director must have stung the lager brewer:

Mr Graesser, we hate lager like a devil hates holy water. For God's sake don't offer us a better lager. I wish it were so poor that everyone coming into our houses would reject it and order our Treble Gold instead.

Wrexham Lager was supplied in small quantities to many British breweries, some of whom, like Hancock's of Cardiff and Fremlins of Kent, began to bottle the brew themselves. Fremlins came to bottle for all Wrexham's London customers. Others bottled in Leeds, Birmingham and Newcastle.

It was as well that Wrexham hung on to the Cunard liner trade, for their catering superintendent, Mr Setterfield, later took up a similar position with the Great Western Railway. 'He was so personally keen on Wrexham Lager that he virtually gave us all the business on the GWR trains and in the railway hotels and refreshment rooms', recalled Bill Graesser-Thomas. 'There was a tremendous business.'

Just before the Second World War, Wrexham Lager made a second major breakthrough into the local trade, when in 1938 it bought all twenty-three houses belonging to Beirnes, after the Wrexham brewery closed down. 'Many were simply rat holes, but there were at least half-a-dozen really good ones in the immediate Wrexham area, and it cost altogether £100,000 to put the really bad ones in working order', recalled Bill Graesser-Thomas.

After fifty-five years of pouring round the globe, the Ace of Clubs lagers at last flowed freely in their home town. The company could even pick and choose whose ales to supply to

BREAKTHROUGH: the brewery made a major breakthrough into the local trade in 1938 when it bought all twenty-three pubs belonging to Beirnes of Wrexham, including the thatched Horse & Jockey in the town, pictured here in 1964 after the takeover by Ind Coope

its pubs. 'We had a stick with which to beat the local brewers, and sales in the Wrexham area went up substantially.'

The war itself proved a mixed blessing. As Wrexham was a garrison town with a large and demanding military population, there was a chance to acclimatize thousands of young soldiers to lager. Trade was brisk. Busy ships' crews were equally eager to escape the tension over a glass or two. Over 85 per cent of production went to the armed services.

Wrexham Lager was even on hand at another moment in history, when Winston Churchill and President Roosevelt met in mid-Atlantic to sign the crucial lease-lend agreement. The wartime deal between Britain and the United States was sealed with a glass of Wrexham draught.

But the ferocious naval war not only meant that Wrexham Lager's extensive export and cruise liner trade was blown away, but also thousands of casks were lost at sea. Brewing materials became scarce, the bottling stores were bombed during the blitz on Liverpool, and the strain began to tell.

The end of the war was celebrated in style with a bottled Victory strong lager sporting a large red 'V' on the label. But all was not well. Chairman Edgar Graesser had a confrontation with sales director Bill Graesser-Thomas over the question of brewing a lower-gravity beer to compete with the cheaper brands like Red Tower from Manchester and Graham's Golden Lager from Scotland.

'I told my uncle Edgar that either he went or I would have to go', said Bill Graesser-Thomas. 'I remember he just walked out of the office in his stubborn way and said if I was going to wreck the brewery, he would have nothing to do with it.' Edgar's brother Norman took over the chairmanship, but his main interest was in Graessers' chemical works.

By 1945, with little opportunity for investment, the brewing plant was run down. The cost of re-equipping proved too great, and after stuggling on for a few years – including the coup of selling their lager on the first peacetime voyage of the famous liner *Queen Elizabeth* in 1946 – the company sold out to established Burton brewers Ind Coope & Allsopp in 1949.

VICTORY LAGER: the Wrexham Lager Company was one of the few breweries to produce a victory beer in 1945. It was still anxious to demonstrate its loyalty to Britain

Ind Coope was busy trying to corner the market in British lager, which, though a tiny sector, it believed had huge potential. At this time lager accounted for no more than 250,000 barrels a year in Britain, including imports. The Wrexham Lager Brewery itself was relatively small, with a capacity of less than 1,000 barrels a week.

In 1951 Ind Coope bought Archibald Arrol's Alloa Brewery in Scotland, adding another famous name, Graham's, to its lager portfolio. It decided to concentrate on the Scottish brand. Wrexham dark lager was discontinued in 1952, but the lighter lager struggled on.

When Ind Coope finally tried to replace the name of Wrexham Lager with Graham's in

Wales in 1957, the population was up in arms. 'It is a real insult to our town', said one licensee. The Licensed Victuallers' Association promised to protest. 'The decision has been received with distaste in all circles', reported the local *Star* newspaper.

Many echoed the words of one drinker:

It is an outrage. The brewery has done a disastrous thing in relinquishing the name of Wrexham Lager. I have travelled two-thirds of the world's surface and have drunk Wrexham Lager in practically every country I have visited.

Even the mayor, Councillor W.H. Evans, was moved to comment: 'It is definitely a loss of prestige for Wrexham. I don't often have a drink, but when I do it is a glass of Wrexham Lager.' The town that had once ignored the brew had taken it to their hearts.

Ind Coope soon dropped the Graham's name as well, and in 1959 introduced a new foreign-sounding lager called Skol. But it never sold well in North Wales. The locals stuck to their own lighter brew, even if it became almost an underground drink.

Ind Coope refused to promote Wrexham Lager, seeing it as an irritant to its development of national brands. It even instructed that it not be sold outside existing outlets, and casually sold off the popular Ace of Clubs trademark in 1963 for a pittance to the Northern Clubs' Federation Brewery of Newcastle. But it did invest in the brewery.

A new brewhouse was commissioned at a cost of £2.5 million, opening in July 1963. The new buildings completely dwarfed the old red-brick brewery, immediately boosting output to 4,500 barrels a week and, with the addition of extra fermentation vessels, to 10,000 barrels. Later expansion in the seventies brought the capacity to 18,000 barrels.

It was claimed to be one of the largest and most modern breweries in Europe, employing 240 workers. Ind Coope was by now part of the giant Allied Breweries, and it used the Wrexham plant as its lager development brewery.

Instead of brewing its own lagers for export, Wrexham now turned to brewing foreign lagers for sale in this country. It pioneered the production of a number of beers brewed under licence from abroad, including Lowenbrau of Munich and Castlemaine XXXX of Australia. It was also the first brewery in Britain to brew American Budweiser, but as Allied failed to clinch the contract, the business went to Watney's.

But Wrexham Lager would not go away, and in 1978 Allied bowed to local pressure and restored the name to the brewery. Sir Alistair Graesser, grandson of the founder, unveiled the new company nameplate. Two years later a new logo, showing the early Bavarian tower, was adopted. Some modest promotion was even put behind Wrexham Lager itself, which was still selling in 370 pubs and clubs in Wales.

LOSING ACE: when the name Wrexham Lager was revived in 1978, the company had to introduce a new logo since the Ace of Clubs device had been sold. Appropriately, it showed the early Bavarian-style buildings

In 1982 the brewery celebrated its centenary by brewing its famous dark lager and Pilsener again to mark the occasion, and bottling them under the Ace of Clubs label thanks to a special dispensation from the Federation Brewery of Newcastle who now own the trademark.

In 1990 Wrexham Lager repaid the renewed faith by winning the brewing equivalent of the World Cup, taking the Championship Trophy for draught lager at the Brewing Industry International Awards in Burton-upon-Trent.

Three years later the Wrexham concern linked up with another well-known lager when parent company Allied Breweries merged with the British interests of a certain Danish brewery to form Carlsberg-Tetley. Robert Graesser would probably have approved.

CLUBBING TOGETHER

If pubs and chapels were uncomfortable neighbours in many Welsh streets, clubs and breweries were just as suspicious of each other. Sometimes their relationship was openly hostile.

The early Friendly Societies from the eighteenth century had usually met in pubs, where their extra trade was welcomed by landlord and brewery alike. But these worthy organizations, offering security against old age, sickness and adversity, tended to be closed, conservative, middle-class bodies. Labourers were rarely welcome. The societies did not seek men without a trade who might be a heavy drain on resources because of their uncertain employment.

Not a single member of the Welsh Lodges of Oddfellows for instance was involved in the rebellion in the coalfields and the iron industry which led to the shooting of twenty-six workers by troops in Merthyr Tydfil in 1831 and the execution of one of the leaders Dic Penderyn. Indeed, Davies and Kissack reveal in their book, *The Inns and Friendly Societies of Monmouth*, that the Druids and Oddfellows offered the services of their entire membership as special constables as the unrest grew in 1830.

So miners and other industrial workers who had flooded into South Wales during the nineteenth century, often into remote, isolated areas, sought their own security in working men's clubs and welfare institutes.

As the brewers began to build up their tied estates of pubs, these clubs were seen as rival outlets, particularly after their numbers rapidly increased following the Sunday Closing (Wales) Act of 1881, which shut pubs on the holy day but allowed clubs to open. Some of the 'clubs' set up at this time were little more than illicit drinking dens (shebeens). Brewers were not amused.

Their anger was sharply expressed by Samuel Brain of the Cardiff brewers when he gave evidence before the Royal Commission looking into the working of the Act in 1889.

I could give you numbers of instances where men have gone into clubs and shebeens, and places of that description, on a Saturday night after they have finished their work, and have not come out of them until Monday, and they spend their wages and mix with low society, and all that sort of thing, which was never the case when they could get their dinner beer (in a pub) on the Sunday and stop at home.

When supplies were restricted during the First World War, it was the clubs that suffered, the brewers giving their own pubs priority. There was also friction over rapidly rising prices.

At a meeting of the South Wales Working Men's Club and Institute Union at Abercarn in December 1916, the following motion was carried:

That in the opinion of this branch the action of the brewers in raising their prices by such a large amount, having regard to the fact that they have been obtaining sums over and above the increase in taxation since November, 1914, is entirely unwarranted and unnecessary.

That critical motion carried the seeds of an immediate post-war development, the beginning of a rare co-operative enterprise in brewing in Wales. Still unable to obtain reliable supplies at acceptable prices, the clubs resolved to secure their own brewery, a sub-committee being appointed in 1918 to investigate the possibility.

A formal motion adopting the policy was carried at a representative meeting of clubs at the Cathays Liberal Club in Cardiff on 28 March 1919, and no time was lost in setting the barrel rolling. In May the brewery sub-committee reported that they were 'now in a position to aquire a brewery on behalf of the South Wales clubs'. Members had offered to subscribe £22,000 towards the scheme.

On 17 June a meeting at the South Wales CIU offices in Taff Street, Pontypridd, decided to take an option on plant and premises at Pontyclun, and a new company, the South Wales and Monmouthshire United Clubs Brewery, was formed, with John Davies of Ferndale as chairman. Two months later he was replaced by Huw Richards of Pontypridd, who was to hold the post for twenty years. Only CIU clubs and their members could be involved, and directors were appointed to represent both the clubs and the individual shareholders.

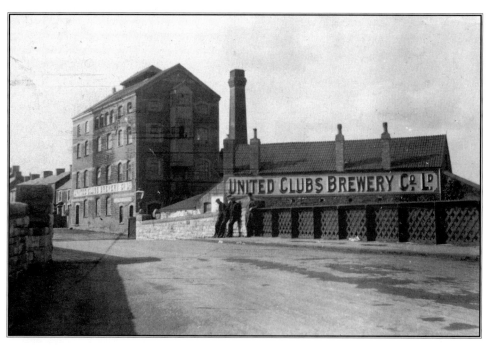

CO-OPERATIVE EFFORT: the United Clubs Brewery shortly after the South Wales clubs took over Jenkins' Crown Brewery at Pontyclun (Crown-Buckley)

The new company began trading in July, with the amount of discount received by the clubs on the price of the beer depending on the size of their shareholding.

The brewery the club collective bought was D. & T. Jenkins' Crown Brewery at Brynsadler, Pontyclun, a tall, stone, turn-of-the-century building built by the bridge on the road from Llantrisant to Cowbridge. Thomas Jenkins was landlord of the Ivor Arms opposite, but the clubmen were not interested in the pubs.

Just £25,000 was needed to form the company and buy the brewery, though a further £15,000 worth of shares was issued the following year to finance expansion. Nearly 500 applications were received: 45 clubs were issued with 16,240 shares while 445 individuals took 23,760.

Jenkins was known for its 'celebrated Crown PA' but the clubs' first head brewer, Captain W. Rogers, a cavalry officer in the Boer War, was initially told to produce another Jenkins' beer, XXXX. This dark mild was described as a 6d. beer and sold at £5 a barrel. CPA or Clubs Pale Ale was not introduced until February 1920, at 4s. more a barrel. In an unusual arrangement, the clubs' brewery initially paid D. & T. Jenkins royalties for these brews.

Production started at around 150 barrels a week, with the board quickly pressing ahead with improvements. Three additional fermenting vessels were added immediately in the brewhouse, 200 casks were bought and Jenkins' horses and carts were replaced by a new 4-ton lorry to complement the existing vehicle.

This road transport was inadequate and soon warnings had to be given not to overload the two lorries. Much of the beer was delivered by rail, especially by the Taff Vale Railway, and some clubs collected their casks themselves from the brewery.

The clubmen were strangers to brewing, as is evident from the board minutes in October 1919, when one of the former owners of the brewery, David Jenkins, was paid to visit Holt Brothers Brewery at Burnham in Somerset to inspect and buy further casks. Later he examined vessels in Banbury for the company.

Not everyone was happy with the beer. Ton Pentre club complained about the quality and was asked to send a sample to the brewer; Mountain Ash WMC reported it had lost 18 gallons owing to a cask blowing out. As always, the clubmen put the matter to the vote.

It was proposed by Mr George Davies, seconded by Mr Dan James, that this club be fully compensated for the loss. An amendment was moved by Mr Howe, seconded by Mr Smith, that the club be compensated to the extent of half the value of the loss. On a vote being taken, the amendment was declared carried.

Though without training, the directors closely supervised the brewing, firing off regular instructions to the brewer. 'Resolved that the secretary write the brewer instructing him to keep a record of all beer returned.' 'Resolved that the secretary be instructed to prepare forms for the brewer to fill up giving particulars of each brew.' Two of the ten directors attended the brewery every week on a rota basis.

In a remarkable admission of short measure, clubs were given an allowance of one barrel in thirty-six to cover wastage and normal shortage. 'In case of an abnormal shortage the secretary of the club to report the same and the barrel to remain undisturbed to enable a representative of the brewery to inspect same.' It was a system open to abuse.

Sub-committees and working committees proliferated, and much time was spent discussing directors' fees and allowances. A building committee was established to consider a bold scheme to build their own offices, club premises and bonded store in Mill Street, Pontypridd, next to the County Hotel. The site alone cost £1,100.

They were worried by the water supply to the brewery 'which owing to the long drought had been giving considerable trouble and keeping down the output'. Again, they sought David Jenkins' advice. And again appointed another working group, the water and boiler sub-committee.

Everything was considered in detail:

Resolved that the premises be painted. Proposed by Mr George Davies, seconded by Mr D. Jones, that the name of the company be substituted wherever the name of the old firm appears. An amendment was moved by Mr A. Pearce, seconded by Mr H. Howe, that the name of the company be painted in one place only. On a vote being taken, four voted for the amendment and five for the proposition. The proposition was declared carried.

The board resolved that the name to be displayed was 'Clubs Union Brewery', rather than the full title. This name was also painted on the lorries. However the Clubs Union quickly objected, causing the brushes to hit the brickwork again. The reworded title came out as 'United Clubs Brewery'.

More embarrassingly for a working men's organization, the brewery workers were not happy with their pay. They downed tools and called in Mr Hall, secretary of the Workers Union, just before Christmas. The board, predictably, responded by appointing a wages sub-committee to meet him. 'The committee were instructed to ask the men to return to work at once with a guarantee that the question of wages would be gone into.'

Feelings were obviously running high, with some sensing betrayal.

A discussion took place as to the discipline existing at the brewery and the control of the men employed, and there was a strong expression of opinion that a foreman should be appointed by the board who was thoroughly in sympathy with the main objects of the movement.

This resulted in one of the directors, T. Rich, being appointed foreman at £5 a week, plus a free season ticket from Cardiff to Pontyclun. At the same time the consideration of an increase in wages was deferred, though members of the wages sub-committee were voted 15s. each for attending. A happier resolution was the decision in the same month to supply 100 barrels of free beer to the clubs for Christmas.

Clubs themselves became involved in the issue. The Crumlin Workmen's Club wrote asking for details of employees' wages. This was refused. And when the deferred issue of workers' pay came up again in January 1920, the board responded – by appointing another wages sub-committee 'to go into the matter'. The immediate dispute was not resolved until February, and not every worker survived the experience. The minutes of 18 February 1920 concluded: 'Resolved that the secretary be instructed to inform Mr David Jones, the dispatch clerk at the brewery, that the directors would be glad if he would endeavour to obtain another appointment.'

Relationships with other breweries were little warmer. When the South Wales Brewers' Association wrote inviting the company to join, the board refused. In fact, it was mounting a campaign attacking conventional breweries.

A poster sent to clubs listed the reasons for taking its beers:

1. Can your club afford to help to pay for the losses on brewery tied houses?
2. Is there any reason why you should not take advantage of the better value offered by the Club Brewery? (Many brewers charge our clubs 5s. or 6s. per barrel more than the Club Brewery prices.)
3. Does your committee pay this higher price for any of your supplies?
4. Can we convince you that our products are quite as good? (Our beers are brewed from pure malt and hops – no chemicals.)

Most clubmen combine in trade unions and co-operative societies to protect their own interests. Why not support your own brewery exclusively, and take full advantage of combined trading?

The poster concluded with a series of snappy slogans:

Try it – Buy it – and don't be talked out of it. Spend where you save. Bonus on purchases to all clubs, whether shareholders or not.

Trade was slowly picking up. The Club Brewery found it necessary to invest in larger casks, swapping its kils (18-gallon casks) for hogsheads (54 gallons) in a two-for-one exchange deal with Hicks Brewery of St Austell in Cornwall. More were bought from further afield from George Younger of Alloa in Scotland. Loans were also beginning to be made to clubs. Ynysddu Workmen's received £100 in April.

Complaints about the quality of the beer continued, however, and in March the full board met the brewer at Pontyclun:

The brewer was questioned particularly as to the gravity of the beer. The brewer then left and after further discussion Mr George Davies proposed, seconded by Mr J. P. Davies, that as from 1st April the brewer be instructed to increase the gravity by one degree (to 1040).

One gains the strong impression that the board members were enjoying their new authority. 'Resolved that a deputation should visit distillers for the purpose of obtaining supplies.' All expenses were paid for the three-man trip to Scotland, plus 30s. a day. On 2 July the national executive of the Clubs Union visited the brewery and was lavishly entertained from the time it arrived in Cardiff.

But the directors were still splashing about in unfamiliar waters. In the summer the simmering wages wrangle heated up again. The board had been considering offering 5s. a week over the standard rate, but during negotiations the workers had walked out, clearly angering the directors. At a board meeting on 28 July the secretary reported that 'the men did not recommence work until 10.30 this morning.'

The directors were in bitter mood and withdrew their increased offer. They went further: 'Resolved that the lost time should be worked back by the employees', and 'Resolved that notice should be given to the employees on Friday.'

A crippling confrontation was only avoided by the action of Mr Hall, the secretary of the Workers' Union, who attended a meeting of directors on 4 August and submitted the terms agreed between the Brewers' Association and other brewery employees. 'After hearing Mr Hall it was resolved that such terms should apply to the employees at the brewery.' The notices given to the workers were withdrawn.

It had been a traumatic first year for all concerned. In the end the brewery had produced 9,780 barrels, an average of 188 a week. This was probably only half of what had been hoped. When the head brewer Mr Rogers had been appointed, he had been granted a commission of 3d. on every barrel produced over 20,800 a year.

But the unique venture had survived. Many other breweries at the time were struggling. In October 1920 it had been offered the chance to buy the Cefn Viaduct Brewery near Merthyr Tydfil, but after inspecting the premises had decided not to proceed. It also pulled out of another expensive venture when it did not to go ahead with its building plans in Mill Street, Pontypridd, instead deciding to sell the site.

Chairman Huw Richards at the company's first AGM at the New Park Liberal Club, Cardiff, on 23 October 1920, could tell the seventy-five delegates present: 'The directors have pleasure in congratulating the company upon the very successful initial year's working.' A profit of £2,265 had been made.

> The new venture has been supported well by the clubs interested, and this support has indicated to the directors the need for a considerable development and extension of the company's plant and buildings. Various plans and schemes for increasing the output are well in hand, and it is hoped that before another 12 months have elapsed, the position of the company will be infinitely better than it is today.

The work in hand included six new fermenting vessels.

There were also plans for its own bottling hall, but this project was repeatedly put off because of lack of sufficient funds, and was not carried out until the early 1930s.

> The loyalty of the individual shareholders, and particularly of the clubs and club committees, has been exceedingly gratifying, and there is no doubt that today this company stands pre-eminent in the country as the most important club union brewery.

The board's lack of expertise was still exposed by transport problems, even though two more lorries had been bought. Casks regularly went missing on the Great Western Railway. Clubs in Monmouthshire complained of delayed deliveries, and eventually decided to set up

THE COMMITTEE: the brewery's ruling body in 1961 under the chairmanship of Bertie Rowe of Pontypridd (Western Mail)

their own brewery, the Gwent Union Clubs Brewery, in an existing brewery at Fleur-de-Lis near Risca.

It was a period when a number of clubs breweries were established in Britain, especially in the Midlands and the north of England. In the spirit of the clubs' movement, the breweries helped each other. A delegation from the proposed West Midlands Clubs Brewery of Willenhall, near Wolverhampton, had visited the Pontyclun brewery in September 1920, and in the autumn of 1921 a Clubs Breweries Federation was formed at a meeting in London.

The Pontyclun company could have bought the Fleur-de-Lis plant and premises itself, but was happy to let the Monmouthshire clubs strike out on their own, and even actively assist them, as was shown by the minutes of 16 March 1921.

> The question of the purchase of the Fleur-de-Lis Brewery, which is now for sale, was discussed. It was finally resolved that if the Monmouthshire clubs wished to purchase it for themselves, this company would lend money to them on the security of their shares in this company.

At least two directors, including the chairman Huw Richards, attended the meeting of the Monmouthshire clubs buying the brewery, and some were involved in both companies. One of the Pontyclun directors, George Davies of Blackwood, became secretary of the new venture, and the account of the Gwent Union was guaranteed to £2,900 by the company. The Pontyclun brewery also sold Fleur-de-Lis various vessels.

This spirit of co-operation did not mean there was no friction. For instance there was the case of the missing pipe which caused much correspondence. The minutes of 15 March 1922 read:

A discussion ensued on the reported shortage of ten feet of piping in the old round purchased by the Gwent brewery and on their request for compensation. It was finally resolved that the company refuse to accept any responsibility.

After more correspondence this decision was overturned on 24 April:

A letter was read from the secretary of the Gwent brewery with reference to the rounds originally purchased from the company. After lengthy discussion it was resolved that this company pay for the ten feet of copper piping which was missing.

The cost was £3.

Despite this sum, the Gwent venture was not a success, and seven years after the brewery was launched, its affairs were wound up. A meeting of shareholders in July 1928 resolved to pass over their shares to the United Clubs Brewery and the clubs involved were again supplied from Pontyclun.

The Pontyclun company did consider buying another brewery in 1924. The secretary of the troubled Cefn Viaduct Brewery attended a board meeting in April when he asked the board 'to consider purchasing the shares in that company with a view to acquiring their trade and business.' Negotiations were entered into, but came to nothing.

The company also diversified into hotels, becoming involved in 1922 in the Langland Bay Hotel, Swansea, for the use of convalescing club members. This move proved time-consuming and unprofitable and the hotel was eventually sold in 1957, after many moves to dispose of it earlier.

The mid-1920s were difficult times, particularly with the General Strike in 1926, when the brewery made many grants to local relief committees and distress funds. That year the bank objected to a dividend being paid to shareholders because of the company's large overdraft.

Slowly, the clubs brewery began to shine, helped by the rapid growth of the CIU in South Wales. In 1921 there were thirty-one affiliated clubs in Cardiff; ten years later there were sixty. At the same time the number of pubs declined. The clubs' trump card was Sunday opening, allowing them to serve for five hours while public houses in Wales were shut. Customers flocked in. By 1930 the brewery was rolling out more than 16,000 barrels a year.

This unfair competition was bitterly resented by brewers and pub landlords alike. A delegate to a meeting of the South Wales Trade Defence League under W. H. Brain in 1927 reflected their feelings:

In the old days people used to attend church and afterwards go straight into a public house for a glass of beer. Today they ignore the churches altogether and spend their Sundays in clubs.

Clubs were regularly invited to visit their brewery, though as a board minute of July 1923 revealed, these happy occasions were not without problems: 'It was felt that there had been some abuse of this privilege.' The directors laid down rules restricting numbers (no more than twenty-five) and frequency of visits (no more than one a year). 'It was further decided that the calling at the brewery by club parties early in the morning when proceeding on an outing should be discouraged.'

The brewery's special method of trading helped its growth. A financial bonus was introduced in 1920 based on the number of barrels bought by a club. As trade increased so these bonuses mounted up. From 1920 to 1968 nearly £3 million was returned to the clubs, allowing them to keep the price of a pint down or extend their premises – and so either way increase beer sales. The brewery's motto became 'Loyalty Pays'.

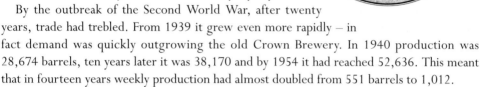

By the outbreak of the Second World War, after twenty years, trade had trebled. From 1939 it grew even more rapidly – in fact demand was quickly outgrowing the old Crown Brewery. In 1940 production was 28,674 barrels, ten years later it was 38,170 and by 1954 it had reached 52,636. This meant that in fourteen years weekly production had almost doubled from 551 barrels to 1,012.

By using every inch of space and pushing the vessels to the limit, the creaking plant could just about cope. Extra vessels were added, but the extra weight was literally causing the brewhouse to bulge at the seams. Tie-bars had to be added to stop the walls collapsing. An engineer's report said that 'in course of time, the foundations will become unsafe'.

The competitive price of beer in the clubs – driven by the success of the clubs brewery – was rattling the established brewers. In 1949 it was reported that clubs were selling beer at 8d. to 9d. a pint compared to 1s. 2d. in pubs. Hancock's chairman J.G. Gaskell threatened to stop supplying clubs if they continued to undercut pub beer prices.

With demand still escalating, the board decided to build a new brewery in fields behind the old one. And, as always in the club world, it did it with a little help from its friends. Clubs in Wales contributed to a new share issue and a sister clubs brewery stepped in to complete the financial package through a generous loan.

Altogether £110,000 was raised through the creation of new share capital, but this was well short of the total cost, so the South Wales board approached the Northern Clubs' Federation Brewery of Newcastle, who agreed a loan of £150,000 at moderate interest terms over ten years. 'Were it not for the sympathetic attitude and the practical approach, both financially and technically, of the directors of the Northern Clubs Federation Brewery, the new brewery as planned would have to have been curtailed', the Pontyclun board recorded in grateful appreciation.

The first turf was cut on 15 October 1951 by the secretary Trevor Williams. The post-war shortage of materials, particularly steel, meant that the building was not completed until February 1954, with the first brew on 9 March. Fittingly, the official opening ceremony in June 1955, was performed by Sydney Lavers, chairman of the Federation Brewery.

The new brewery was capable of producing 2,600 barrels a week. Head brewer Lee Marsh from a Dorset brewing family had taken over when Mr Rogers died in 1936, and supervised all the problems of moving plant from the old brewery to the new building. After the initial teething troubles were overcome, he introduced in 1960 a new draught beer called SBB (Special Best Bitter) to add to the popular CPA and 4X.

UNITED ACHIEVEMENT: the modern brewery at Pontyclun built in the early 1950s with the help of the Northern Clubs Federation Brewery of Newcastle. Tank beer delivered by road tankers was introduced in 1963

This new brew not only went on to account for a third of the company's output, it also took the championship for the best cask-conditioned draught beer at the Brewers' Exhibition in London in 1964. Twenty years later Crown's Brenin Bitter took the prize for the best brewery-conditioned beer at Brewex.

This second award reflected the switch in brewing over the period from traditional cask ales to mainly processed beers. First the brewery in 1963 introduced filtered tank beer, delivered to the clubs by large road tankers and then piped directly into tanks installed in the cellars. This cut the amount of labour required for handling large amounts of beer in the big social clubs.

For smaller outlets pasteurized keg beer was added in 1969. This proved popular as many bars in places like rugby clubs had variable trade and inexperienced staff. The leading brand was Great Western; another was called by the strange name of 'Same Again'.

By 1969 a fleet of twenty lorries was serving three hundred clubs scattered over industrial South Wales from Coleford in the east to Carmarthen in the west and bounded by Brecon to the north. A western depot was later established in Narberth.

Much of this trade was sealed by loans. In the 1960s the brewery loaned their customers the formidable sum of £2 million. This paid for lavish improvements in the clubs while guaranteeing the brewery's products on the bar. To help finance this system, a further 150,000 shares were issued in 1961, bringing the share capital to £300,000.

The bonus scheme was also still going strong. In 1964 the Bay View Social Club of Port

Talbot downed 900,000 pints of clubs beer. For this tremendous thirst it earned £6,684 in bonus, the best that year. Altogether in 1964, £234,000 was shared between 240 clubs. 'In the last five years we have given £790,000 back to the clubs', said secretary Penri Evans, who was to become the company's first managing director in 1967. 'All the clubs have used the money wisely and some of them are a treat to visit.'

Clubs like Tonyrefail Non-Political were typical. It had received £12,000 in bonuses in the previous five years, and in 1964 gained a further £3,000. In the same year it opened a new extension, and nearly the whole cost was covered by bonus payments.

The company was firmly rooted in its local community, employing 135 workers, most from within a five-mile radius of the brewery. Many had long connections, seven between them boasting 234 years service. The Rosser family of Llanharry had six brothers employed there. There was even a waiting list for jobs. This was hardly surprising since by now the clubs brewery paid substantially higher than standard union rates. Once in, few ever left. This stability was also evident in the senior posts. In its first fifty years the company had only three secretaries, three head brewers and five chairmen.

Some of the benefits of working there were less obvious. A stream, the Nant Felin Fach, ran by the brewery and was for many years used to dispose of waste materials like excess yeast, cask washings and spent hops. This murky soup bred worms and other aquatic creatures, and these in turn attracted shoals of trout. Workers found that night lines set from the brewery were almost always successful. When modern regulations demanded different disposal methods, the fish stopped biting.

The mid-1970s saw two major changes. In 1976 the exclusive link with the clubs was broken, and the brewery began selling to a wider free trade. In 1977 the cumbersome name of South Wales and Monmouthshire United Clubs Brewery was altered to Crown Brewery, a title which could be much more easily swallowed by their new pub customers.

By then the company was supplying over 500 clubs not only in South Wales, but also in Birmingham and the west of England, with some 2,500 barrels of beer a week. Not all of these were brewed at Pontyclun, lager for instance being bought in. The Crown Brewery itself produced 1,700 barrels a week – 25 million pints a year.

Traditional cask-conditioned ale, which accounted for a fifth of production, was remarkably still delivered in wooden casks of Polish oak (most breweries had by now switched to metal containers). And to keep these in good order, Crown employed a cooper, one of the few left working in Wales.

Despite its growth in size, the personal touch still remained important. Instead of advertising campaigns to sell its beers, Crown continued to believe in hospitality. Club members were regularly invited to the brewery. On average three visits a week were arranged. The sample room was one of the most popular bars in Wales, with visitors enjoying 75,000 free pints in 1977.

In 1984 the brewery widened its horizons further by setting up a depot in London and then at the end of the year opening its first pub, the St Oswalds in Port Talbot, in converted club premises. More pubs quickly followed.

Slowly the remarkable venture was changing into a more conventional brewery, with its own tied estate. In 1989 it was to take a further major step in this direction when it linked up with one of Wales's oldest breweries.

CONCENTRATED ALE

If Wales had arrived late at the big brewing party, it also left early. When *The Times* produced a book in 1959, *Beer in Britain*, the accompanying map was so packed with breweries, it was almost impossible to read. London still had 22 companies, the Manchester area 20, Birmingham and the Black Country 15 and Edinburgh 18.

Yet in Wales there were wide open spaces. Wrexham, the Burton of Wales, could only offer two breweries – Border and Wrexham Lager. Even in the South Wales coalfields and industrial towns, which had once been flooded with local beer, only a few firms were left. Swansea boasted just one remaining brewery, Hancock's. Merthyr Tydfil had none at all.

Terry Campbell, writing in the *South Wales Echo* in 1977, lamented the loss:

> Just look at the breweries South Wales has lost over the years . . . There were the Taff Vale and Giles & Harrap which kept the Merthyr Tydfil areas bubbling. Over the mountain the Black Lion slaked the thirsts of Aberdare. Then we had Evan Evans Bevan rolling out the barrels at Neath and in Mid-Glamorgan the Bridgend Brewery was kept busy keeping everyone's pints foaming. Also in a glass of their own at Cardiff were the redoubtable Hancock's and Ely breweries. Sparkling up at Aberbeeg were Webbs, while Phillips and Lloyd & Yorath of Newport kept a cool head in South Gwent. Now Brains of Cardiff, Felinfoel Ales and Buckleys of Llanelli are the sole remaining independent brews left in South Wales.

After mergers within Wales most of the lost brews had fallen to major English companies. In the nineteenth century breweries from Burton, Birmingham and London had sold their beers into the rapidly-expanding Welsh market, establishing a considerable trade especially in South Wales. In the twentieth century they had followed up by taking over local companies. Allsopps of Burton had wrapped up the leading breweries in Aberdare before the First World War; Trumans of London had captured Swansea United Breweries in 1926.

But it was after the Second World War that the froth of Welsh breweries was really wiped away. In Newport all the local brews vanished in three years around 1950. In 1949 Simonds of Reading (later Courage) snapped up Phillips and two Birmingham firms rolled up the rest. Mitchells & Butlers took over Thatchers in the same year and Ansells acquired Lloyds in 1951. Some brewing continued for a while, but it was only to produce the larger brewers' brands. The local beers disappeared.

It was all one-way traffic. Unlike Scotland – where William Younger and McEwan of Edinburgh had combined in 1931 to form Scottish Brewers (later Scottish and Newcastle Breweries) to push into England – no Welsh brewing group emerged to develop any significant trade across the border. Hopes that this could happen on a large scale ended in the 1960s.

IN RUINS: the head office of once proud Phillips and Sons of Newport during demolition (Courage Archive)

Hancock's of Cardiff and Swansea, who had looked the most likely to lead Welsh ales into England, fell to Bass Charrington in 1968. A subsidiary Bass company called Welsh Brewers was formed to combine Hancock's with Webbs of Aberbeeg. The other main brewing force in South Wales, Rhymney Breweries, which in 1959 had taken over Ely Brewery of Cardiff, was itself bought by London brewers Whitbread in 1966.

Neither takeover was a surprise in a decade that saw the aggressive growth of Britain's brewing giants – the 'Big Six' of Allied Breweries (Ansells, Tetley and Ind Coope), Bass Charrington, Courage, Watney, Whitbread and Scottish & Newcastle. Nor was it a shock that it was Bass and Whitbread who moved in to dominate the Welsh valleys; both had been closely connected with Hancock's and Rhymney for many years before the eventual takeovers.

Rhymney's final years as an independent company show the mounting pressure on even the largest Welsh brewery. Whitbread's influence was already considerable, with two directors on the board. Col. Harry Llewellyn was even invited to become chairman in 1958 by Whitbread's boss Bill Whitbread.

Col. Llewellyn might be best remembered for winning a show-jumping gold medal on Foxhunter at the 1952 Olympic Games in Helsinki, but he was also an experienced industrialist. Until the nationalization of the coal industry in 1947, he had owned a number of colliery companies, and in 1954 bought the Abergavenny engineering firm of Davis & Co.

STILL STEAMING: Hancock's old brewery in Cardiff, still in operation for Welsh Brewers as Bass's only brewery in Wales

In 1945 he had been decorated by General Omar Bradley of the American 12th Army for the part he had played in planning 'D' Day and the battle in Europe as Field Marshall Montgomery's liaison officer.

A profile in 1958 describes him as 'a restless, hard-driving man who marries a great flair for administration to a very sound knowledge of modern production problems and a steel-hard determination to build his own business house'. When he leaped into the saddle of the hobby-horse brewers of Rhymney, he still wanted to be a winner. His takeover of Ely Brewery within months of taking charge shows the extent of his ambition. If anyone could lead a late cavalry charge for Welsh beer, he was the man.

He started with typical energy. The two neighbouring breweries at Ely (Ely Brewery and Crosswells) were merged with a new £500,000 brewery virtually built on the Crosswell's site in 1962, doubling the potential output from Cardiff. In a period of growing keg beer sales, he marketed two brands, Rhymney's Hobby-Horse and Ely's Silver Drum. He also saw the potential for lager, one of his favourite drinks, at a time when few regional brewers thought there was much mileage in producing this beer in Britain. The yeast was regularly flown over from Pripps Brewery in Sweden to glide out Ski Lager in the mid-1960s.

Like Lazarus Nidditch before him, he thought big – even to the extent in 1962 of erecting a massive £1,000 sign outside the brewery in Ely alongside the main A48 road to West Wales. This grand structure featured a huge barrel pouring beer into a seven-foot high perspex tankard on a tray. The outsize mug took eight-and-a-half minutes to fill, with the 'beer' (an electrically pumped brown chemical) then returning to the barrel in an endless cycle. Alongside were two giant bottles of Welsh Brown and Welsh Bitter. At night the glistening glass was lit up by fluorescent lamps.

Early in 1961 Rhymney took over South Wales's oldest mineral water manufacturers, Hansard & Sons of Merthyr Tydfil, and in June the company was reporting record annual profits of over £500,000. The 700-pub concern also moved into hotels, buying the well-known Angel in Cardiff and the Marine and Seabank in Porthcawl in 1962, besides opening the Dolphin in Swansea in conjuction with Fortes. There were plans for a motel between Newport and Cardiff.

In his annual report in August 1962 Harry Llewellyn was confident about the future. Profits were up and a wide range of new projects were in hand. He had no worry about his relationship with Whitbread. 'We continue to enjoy a close, friendly

SHOOTING IN FRONT: Rhymney was slopes ahead of many of its rivals when it introduced Ski Lager in the mid-1960s

and profitable association with Whitbread, which enables us to offer a fine range of national bottled beers as well as our own.' Bottling of Whitbread's Mackeson Stout helped maintain the profitability of its bottling line.

Even when Rhymney shares jumped later that month, he was unconcerned about a possible bid. 'We don't expect a takeover approach from anywhere', he told the *London Evening Standard*. 'We are a Whitbread "Umbrella" company [i.e. one in which Whitbread held a share stake] and our relations with them are both close and good. I don't see Whitbread wanting a bigger stake.'

Rhymney in fact attempted a takeover themselves, trying to buy Buckley's Brewery of Llanelli. The bid failed but resulted in 1963 in an exchange of directors, Captain Mason Scott of Rhymney joining Buckley's board while the West Wales company's chairman, Col. Kemmis Buckley, joined Rhymney. 'There is, of course, no question of the complete independence of either company being affected in any way, but this interchange of directors does cement the traditional friendship between our two companies', said Col. Llewellyn. Whitbread was closely involved in the arrangement, having a large share stake in both firms. Captain Scott was a Whitbread director.

Rhymney was a leading innovator. Following the success of its keg beers, Rhymney was one of the first to introduce tank beer, whereby road tankers delivered filtered beer by pipe into large temperature-controlled vessels in the pub cellar. Some sixty pubs were fitted out with tanks in 1963. 'The old beer barrel is on its way out', said a spokesman about the easier delivery system that did away with manhandling casks. 'This is another development which may become commonplace in the future', he added.

Vessels of a larger size hit the headlines in August 1963 when two huge 400-barrel vats were transported by road tanker from Leeds to Cardiff for the brewery in Ely. The load – 16 ft 6 in wide by 17 ft 3 in high – was said to be one of the largest ever moved in one piece. Special police permission had to be obtained for the journey, and when the 32-ton

FLYING HIGH: an aerial view of the Rhymney Brewery at its peak around 1960, with an impressive stack of barrels

fermenting vessels arrived they only fitted through the brewery gates 'with an inch to spare'. Once in place an extension to the brewery was built around them.

Rhymney was growing. In 1964 the Newport Wine Lodge was taken over with eight off-licences, and a million-pound share issue was made to fund expansion. Trading profits were £730,000 on business worth £7.75 million. But behind the bold plans cracks were beginning to appear, not helped by the death of the experienced vice-chairman Mr B.W. Kemp-Welch. Suddenly Col. Llewellyn did not sound so confident in his report to the 1964 AGM:

> Competition in South Wales is increasingly keen. Companies such as ours with a high percentage of free trade outlets and a predominantly draught beer trade, upon which they have come to depend for the largest part of their income, are subject to intense rivalry. This comes from local, regional and national brewers who see Wales as an expanding area and are keen to share in its prosperity. There is, however, a limit to the rate of expansion in any area and the rising pressure of competition in South Wales makes it difficult merely to maintain one's sales figures in certain districts.

When he was talking about the free trade, he meant clubs. More than most areas in Britain, clubs in Wales accounted for a larger amount of the beer business. This proliferation of clubs had been stimulated by the Sunday closing law, which allowed clubs to sell drink while pubs were shut. Competition for their trade was razor-sharp. Clubs expected keen prices and other inducements like low-interest loans to fund extensions and refurbishments of their premises. Some Welsh breweries neglected their own pubs in order to meet the clubs' demands. Col. Llewellyn reflected this dilemma.

The workingmen's club like the public house is changing and improving. Some of the most difficult decisions we have to take are those involving demands upon our resources of these two types of retail outlets. Whereas we naturally wish to keep up as high a standard as we can in the condition of our houses and build new houses in suitable places, we have to be realistic in our attitude towards the club, and ready to invest in that direction as well. In certain districts where we rely mainly on our tied outlets, and where their value is seriously threatened by extensive investment in clubs by our competitors, it may well be advisable for us to change our policy and concentrate mainly on the improvement of clubs and the financing of some new ones.

To compete in this league, Rhymney needed extensive funds. Col. Llewellyn was beginning to feel out of his depth. During the year eighteen pubs were lost, a third of them closed because 'they were not capable of improvement at a reasonable cost'.

Rhymney also needed money as they tried to spread their business out of the depressed valleys and South Wales coalfields – where declining demand had cut the Saturday morning shift and 'much of the beer money in the miners' pockets' – and into the more affluent coastal strip.

Whitbread sensed its opportunity. The company's minute book noted in July 1965 that Bill Whitbread had arranged a meeting with Harry Llewellyn 'to discuss the financial situation'. Whitbread held a 30 per cent stake in Rhymney, after an association lasting fifteen years. Negotiations went on for six months.

In January 1966 it bid for complete control. Rhymney's board supported the £4.2 million offer and recommended other shareholders to accept. 'The directors feel that in order to fully exploit the many opportunities now available in South Wales, the company must be integrated more closely with Whitbread as national brewers.'

Not all shareholders were impressed. Arthur Smith, who had sold his Newport Wine Lodge firm to Rhymney, felt the offer was far from generous. 'I wouldn't accept that for my shares.' He hoped for a rival bid. 'I believe that there are other breweries interested in Rhymney.' Another described the merger as 'a marriage with every sign of a shotgun in the background'.

However, once influential banker Julian Hodge accepted on behalf of his substantial holding, there was no prospect of a shareholders' revolt. Alan Thomas, investment director of the Hodge Group commented: 'The acquisition price is poor in terms of asset values, but the price is probably fair when related to Rhymney's likely profit levels over the next few years. Rhymney have not enjoyed the best of circumstances in recent years.'

A spokesman for Rhymney's financial advisors tellingly added, 'I admit the price does not look too generous, but you must look at it in the light of what has happened to this company in the past 18 months.' Profits had stagnated and many commented on the poor state of Rhymney's houses. The hobby-horse felt it could no longer compete in the new super race.

Rationalizations followed. Rhymney's keg beers were dropped and replaced by Whitbread Trophy and Tankard. Llanfoist mineral waters were discontinued. After Whitbread's takeover of Evan Evans Bevan of Neath in the following year, the two were merged in 1969 into a new company called Whitbread Wales under the chairmanship of Harry Llewellyn. The popular hobby-horse symbol was put out to grass, being replaced by Whitbread's hind's head.

MODERN MONUMENT: Whitbread's Magor Brewery built in 1979, sheltering behind Whitbread's hind's head symbol

Brewing ceased at Neath in 1970, but Rhymney Brewery managed to linger on until 1978, despite its awkward position at the top of a valley and a threatening fire in 1968. The Cardiff brewery at Ely brewed its last in 1982. Both were demolished. All signs of the company's Welsh roots gradually disappeared except for the introduction of a low-gravity brew called Welsh Bitter.

Whitbread did, however, choose to build a major new brewery in Wales at Magor alongside the M4 motorway between Chepstow and Newport. The site was primarily chosen because of its vast water supply. When the Severn Railway Tunnel was being built in 1879 it was flooded by the Great Spring which has a minimum flow of 9 million gallons a day, almost doubling in peak periods. This unwanted water is pumped out by British Rail at Sudbrook and Whitbread tapped into this great source.

Magor was built in 1979 at a cost of £51 million, being carefully blended into the countryside across 58 acres. It was not intended to replace Rhymney and Ely, but was primarily designed as a lager brewery. The first brew in September 1979 was for Heineken, but later it came to brew ales as well including Welsh Bitter. Annual production by 1992 was 1.6 million barrels.

Magor might be the most modern brewery in Wales, but the company which stole most of the headlines in the 1980s liked to claim it was the oldest brewery in Wales. Buckley's of Llanelli did not, however, enjoy all the attention and probably wished the journalists were occupied elsewhere. For most of the news was unwelcome.

Colonel Kemmis Buckley, the great, great grandson of the Reverend James Buckley, had taken over the chairmanship of the family firm in 1972 from his father W.H. Buckley. He believed that in part the company owed its continued existence to its distant location. 'I often think that the main reason for our survival lies in the geographical position of our trading area. This is far enough to the West of Wales to have deterred large predator brewing companies in time past from making bids for us.' In the 1980s Buckley's was to discover that there were no longer any safe hiding places.

In 1983 Kemmis Buckley retired as chairman, having overseen a major redevelopment of the brewery in the early 1970s. Control passed outside the family, with Griffith Philipps, a partner in Cardiff stockbrokers Lyddon & Co., taking over as chairman. Colin Thomas was managing director.

It was not an easy time to step into the hot seat. Profits fell by 21 per cent in 1983 to £870,000 and slipped again in 1984 – mainly due to the cost of heavy borrowing to fund the building of new pubs and the refurbishment of their existing 160 houses. Worse, the large Whitbread holding that had once looked so reassuring, was now suspect.

Whitbread had helped Marstons take over Border Brewery of Wrexham in 1984 and analysts suspected that the giant group might be engineering a similar fate for Buckley's. Speculation mounted, but Griff Philipps was confident, relying on the company's remote location to protect it. 'We're rather at the end of the line here.'

Distance was no barrier to Nazmu Virani, a Ugandan Asian who ran a property and finance company, Control Securities. He had developed a taste for beer in out-of-the-way places, having taken control of Belhaven Brewery at Dunbar in Scotland. In 1985 he bought a 6 per cent stake in Buckley's, now Wales's last publicly-quoted brewery, and asked for a meeting with the board. According to the *Western Mail*, he was 'met with looks of calculated indifference'. Profits were up again, with sales reaching £12 million.

Rebuffed, Virani passed on his share stake to a more aggressive player, Tony Cole of Bestwood, another property and financial services group. He splashed out on more Buckley's shares. By November 1986 he owned over 21 per cent of the company and was demanding a seat on the board. Buckley's responded by asking what he could contribute to the firm. Angry, Cole felt this was 'impertinent'. He increased his stake to 27 per cent, but Whitbread matched his move, upping its holding to a similar figure. Control of Buckley's increasingly depended on these two grappling groups. The board seemed powerless.

A showdown shareholders' meeting in February 1987 saw Cole's proposals defeated. 'Now we can go back to work making beer and not have our time wasted by extraordinary general meetings', said chairman Griff Philipps. But the circling predators would not go away. Thwarted by Whitbread's blocking tactics Cole sold his stake to another two stock-market players, who adopted a more softly, softly approach.

The new stake holders were financiers Peter Clowes and 25-year-old Guy Von Cramer, who bought the shares through their Brodian company. They did not demand an immediate seat on the board, but instead held an 'amicable' meeting with the board. Then on the day of the AGM in July 1987 they surprised Buckley's with a full bid for the company worth £26.6 million.

The board rejected this as 'unwelcome and inadequate'. The former description was certainly true, but the latter looked increasingly inaccurate, especially once Clowes and Cramer increased their offer to £28 million in August. The brewery and pubs had been valued at less than £14 million in 1986. The offer looked more than generous, though doubts were expressed about Brodian's ability to finance it. When the offer was upped again to £29 million in September even Buckley's board felt bound to recommend it to its shareholders. Brodian was home and dry.

But if Buckley's thought it had found a safe home it was mistaken. Clowes and Cramer might sound plausible and look wealthy. Guy Von Cramer, who became managing director, liked to turn up to meetings in a Bentley or by helicopter. He even breezed into Felinfoel Brewery assuming that Buckley's 49.5 per cent stake in the private family company gave him control. The Lewis family quickly showed him the door.

Clowes and Cramer's reign at Buckley's lasted little longer. They moved into shop-fitting, buying firms in Yorkshire, and planned to launch a new beer called Blacksmiths, only to discover after all the publicity material was produced that the accompanying anvil trademark was already owned by Hyde's Brewery of Manchester.

More seriously, Clowes's financial empire was seriously flawed. In June 1988 eight months after taking over, Peter Clowes announced that he was selling his 40 per cent stake in Buckley's. The news followed the Securities and Investment Board's liquidation order against his Barlow Clowes gilt management company. When they defaulted on the loan made to them by Singer and Friedlander to take over the brewery, the merchant bankers moved in and Cramer was forced to resign.

Corporate troubleshooter Morgan Grenfell was called in to sort out the mess. Buckley's shares were suspended for ten weeks and a loss of £763,000 was recorded for the last nine months of 1987. Temporary chairman Michael Willcocks admitted that the closure of the brewery was a distinct possibility.

NEW FORCE: *Crown-Buckley brought together two very different breweries*

Then in November 1988 a black knight rode to the rescue. Guinness, through their Harp Lager subsidiary, bought up Clowes and Cramer's 53 per cent stake at 156p a share – 36p below the takeover bid price of 192p. The deal also involved the Club Brewery, Crown of Pontyclun, who bought the brewery from Harp and formed a joint company with them to run the pubs.

The new Crown-Buckley venture, officially launched in 1989, meant the end of brewing at Pontyclun as production was concentrated at Llanelli. Heavy promotion was put behind Buckley's Best Bitter, which was seen as the new company's flagship brand, but trade proved difficult in a fiercely competitive market.

In 1990 Guinness took a controlling interest in Crown-Buckley, gaining 75 per cent of the company's enlarged share capital through the conversion of substantial debts owed to Harp. The major brewer took charge, with three of the five directors on the board.

The move was vital as Crown-Buckley lost £1,187,000 in 1989 followed by £1,076,000 in 1990. 'In January 1990 the company became unable to meet its commitments to its bank and creditors', stated the annual report, causing Guinness to step in. To look after its interests the Dublin brewer recruited Mike Salter, a former finance director with Bass in Wales, to take over as managing director from Gareth Thomas.

'There was a grave financial crisis. Without the injection of cash from Guinness the company would have folded', admitted Mike Salter. 'Crown's club trade was in decline and Buckley's had lost a lot of business during their troubled years. The business had suffered one blow after another. I think most of the staff and many customers were left shell-shocked by it all.'

The constant changes were not yet over. In 1993 Mike Salter negotiated a management buy-out of Crown-Buckley from Guinness, restoring the company as an independent Welsh brewery.

VANISHED VENTURES: beer mats and labels from eight Welsh breweries which briefly sprang into life and then disappeared

PINT SIGHS

The constant contraction of the brewing industry in Wales and the rest of Britain was thrown into reverse in the late 1970s and early 1980s, when a pint-size revolution stirred the bars, delighted drinkers and surprised the major brewers.

Around 300 new breweries sprang into life across Great Britain, some just brewing for one pub alone, others developing a widespread free trade and a few buying their own handfuls of houses. They might be tiny, no more than a few grains of malt in Britain's mash tun, accounting in total for less than 1 per cent of the beer brewed in these islands, but they added welcome variety at the bar and even pioneered the resurgence of some neglected beer styles like porter and draught strong ales.

In Wales they put a smile back on the business face of brewing with beers like Master Blaster (Afan Brewery), Druid's Ale (Gwent Ales), Piston Bitter (Monmouth Fine Ales), Snowdon Strong (Gwynedd Brewers) and Son of a Bitch (Bullmastiff Brewery).

But brewing in Britain was by then a rock hard market to enter. Most pub doors were already shut firmly in the face of this brave new world of beer. The national companies and established regional firms either owned most of the pubs or had tied up the so-called 'free trade' through cheap loans.

By the 1990s over 200 of these pint-size pioneers had gone to the brewhouse wall, unable to find sufficient outlets to sell their beer. In Wales the casualty rate was even higher. Out of a total of eighteen new breweries set up in the Principality, only two were still rolling out the barrels by 1991.

The initial heady froth and ferment was aroused by the growth of a vocal consumer movement, CAMRA, the Campaign for Real Ale, in the early 1970s, following the emergence of the 'Big Six' national combines of Allied Breweries, Bass, Courage, Scottish & Newcastle, Watney and Whitbread in the previous decade.

This spontaneous consumer movement, at first dismissed by the national companies, not only breathed new life into the remaining independent breweries and eventually persuaded the giants of the industry to reintroduce traditional cask beer – it also sparked a small brewery revolution.

Inevitably some of the early campaigners put their money where their mouths were. Martin Sykes, a member of the first CAMRA National Executive, opened the first new brewing company in Britain for fifty years – Selby Brewery in Yorkshire in 1972. 'I foresaw the revival of interest in real ale, and got in early', he explained.

Ventures followed at the Miners Arms restaurant at Priddy in Somerset in 1973 and at the Mason's Arms pub in South Leigh, Oxfordshire, in 1974, reviving the ancient art of the home-brew house, which by then was in serious danger of extinction. Only four remained in Britain, including one on the Welsh border, the historic Three Tuns at Bishop's Castle.

The first tentative steps in Wales were taken by the Miskin Arms at Miskin, near Pontyclun, which sold its own malt extract brew for three years from 1976. Another licensee, Dafydd Gittins of the Camden Arms in Brecon, experimented with setting up his own Brecon Brewery in an old brewhouse behind the pub in 1978–9, to produce a beer called Black Dragon. But he found the spirit more willing and decided to concentrate on his successful Welsh Whisky business. After blending Scotch whisky for a number of years, he opened his own distillery in Brecon in 1993.

More persevering was Richard Hall of the Globe Inn in Fishguard, who began brewing Black Fox Bitter in 1981. 'Customers cheered loudly when in reply to the Chancellor's 5p on a pint duty levied in April, 1981, I said I would brew my own beer and take the 5p off.' Regular Wynford Vaughan-Thomas pulled the first pint of the price-cutting brew.

But would such beers survive outside the protection of their own pub, in the fierce free trade? Keen home-brewers Bob Parker and Gareth Lewis were among the first to try and answer the question.

They established Bragdy'r Defaid Du, the Black Sheep Brewery, in a disused prefabricated building overlooking the River Teifi, near Newcastle Emlyn, in 1980. They adapted a big copper tank to heat water and made their own mash tun to brew 50 gallons a week for Lyn Jones of the Alltyrodyn Arms, 15 miles away in Rhydowen.

But they did not take the plunge into full-time brewing, keeping their regular jobs while working flat out at weekends on their Black Sheep brew, with its attendant VAT and Customs & Excise paperwork. After two years they decided the trial was not a viable proposition.

BLAST OFF: John McCardle (left) and Ian Barber of Afan Brewery, Port Talbot, celebrate the launch of their strong ale, Master Blaster, in December 1982 (Western Mail)

But others were less cautious. By 1982 there were eight new breweries steaming away in Wales, reaching a peak of nine the following year. It was a brief period of enterprise and exhilaration. But it was not easy.

Afan Brewery of Port Talbot was almost frozen out before it poured the first pint in 1982. 'We began brewing on January 1, but it was so cold the beer never worked', said founder John McCardle. 'Then we were snowed in, and local landlords came and asked us for some beer, and we had to turn them away. It was heartbreaking, especially since we had hoped to be in production by Christmas.'

The company made history when it finally rolled out the first barrel of Afan Bitter by the end of the month from a 15-barrel plant in an industrial unit on the Brunel estate at Cwmafan. 'For there has never been a brewery in Port Talbot before, owing to restrictive convenants in the past preventing the production of alcohol', claimed John McCardle, who with partners Ian Barber and John Grimmer, sank £40,000 into the venture.

Red Kite of Aberystwyth flew into problems before it even reached the brewing stage. 'I think the most frustrating time for us was when we were waiting for a selective financial assistance grant from the Welsh Office', said founder Kieran Healy. 'We were told a decision would be made in eight weeks, but it took them about 18 weeks to tell us we would not be getting the grant.'

Then the Development Board for Rural Wales, who owned the brewery building at Glanyrafon, took another nine weeks to re-assess the position. 'We were kept waiting a total of 27 weeks, when all the time we were ready to go', said Mr Healy. 'We researched the whole situation very thoroughly and found that this area was crying out for its own real ale.'

In 1980 in North Wales four real ale drinkers started dreaming over their pints in the Albion pub in Chester, led by Peter Clay who had just left his job as a purser on a cruise liner with a golden handshake.

> I thought, here are four people with various talents and a common interest in drinking. The more we talked about starting a brewery, the less we laughed. We realised it was not just a pipe dream — it was possible. Between the four of us we knew friends, and friends of friends, who would be useful to contact. The first was a master brewer.

The youngest of the four, Phil Lightfoot, had gone to school with the son of Peter Austin, one time head brewer of the Hull Brewery. By the early 1980s, Mr Austin had not only established his own successful Ringwood Brewery in Hampshire but also helped set up a number of other small breweries around Britain and abroad, supplying both equipment and expertise.

With his aid, Cestrian Brewers rolled out its first barrel in 1981 at CAMRA's Wrexham Beer Festival from a site on the Pinfold Lane estate at Buckley in Clwyd. 'Our research has revealed that there is a local market for a distinctive and different pint, and we hope that Cestrian will have a slightly dry but hoppier flavour', said another partner Geoff Terry.

But within little more than two years, sometimes less, ventures like these had folded. Cestrian and Red Kite closed by the end of 1983; Afan in 1984. It was not through lack of expertise. Peter Austin advised both Afan and Cestrian; Red Kite had the assistance of another experienced brewer, Rory Garden, who worked for the plant suppliers Bruwell of London.

Another small Welsh concern, Powys Brewery of Newtown, was set up in 1981 by a professional brewer, Stuart Roberts, who had previously worked for Allied Breweries at Wrexham. His wife, Elizabeth, had been a research micro-biologist for Scottish & Newcastle. Yet the company shut the following year.

Their beer was good. Their common problem was finding sufficient outlets to sell their ale. In the towns and larger villages, the major brewers owned most of the pubs. When Bullmastiff Brewery set up in Penarth in 1987, it could find only one bar in the whole of Cardiff to sell their beer.

In contrast, the city of Bristol and surrounding area boasted a large number of free houses, allowing successful new companies like Smiles and Butcombe Brewery to thrive. Today their beers are even sold into south-east Wales.

In the scattered rural areas of Wales, where free houses were more plentiful, many were tied to the major brewers through cheap loans. In the others competition for space on the bar was fierce, and trade was often only busy during the short-lived holiday season.

One obvious answer for the new brewers would have been to buy their own pubs. But often having mortgaged their future just to buy the equipment and casks, most lacked the financial muscle necessary to invest in their own houses, let alone compete with the cheque-book marketing of the major firms.

There was certainly nothing wrong with the concept of brewing on a small scale — as was proved by the major brewers copying the idea. In London David Bruce's chain of eccentric but highly successful 'Firkin' home-brew houses led to a host of imitators. Whitbread set up a number of its own home-brew pubs around Britain, including one in 1983 in Cardiff. But the siting of the Heritage Brewery in a suburban house in St Mellons, instead of the city centre, doomed the project.

Allied Breweries preferred to think on a larger scale, building a brewery behind the City Arms at Minera, near Wrexham, in 1983, not only to serve the pub but also a number of their other Lloyd & Trouncer houses in the area. Minera Bitter proved successful, and brewing was only discontinued six years later after reorganization within the Allied group.

A few independent new breweries started promisingly. Gwynedd Brewers of Anglesey, who began fermenting in 1980 at Gaerwen, were helped by the fact that the founding Wainwright brothers owned a small chain of hotels in North Wales. When Cwrw Mon (Anglesey Ale) was launched at the island's annual show, the future looked bright. Four years later Gwynedd had stopped brewing.

Director Owen Wainwright had no doubt about the reason why. 'The big brewers have got all the free trade sewn up through loans. There was no prospect of increasing trade, and we were just not making money.' In addition he blamed bad debts, which were a much more crippling problem for small firms. 'We lost a lot of money in four years. The last straw came when a college refused to pay a major debt of £2,500.'

Phil Silverthorne's Gwent Ales battled for a similar period, first from an old brickworks at Little Mill, near Usk, and then from Cwmbran, before closing in 1985. A former engineer, he produced one of the widest ranges of any new brewer including Wales's strongest draught beer, Druid's Ale (1072). Some fifty pubs and clubs took the various brews.

The company had many bold ideas but was hard hit by planning decisions. An imaginative scheme to convert a disused Co-op store in Pontnewydd into a Victorian home-brew pub was rejected in 1983; the year before an application was turned down to open an off-licence in New Inn near Pontypool selling beer in four-pint jugs.

Phil Silverthorne helped Trevor Buffery and Alan Statham set up a home-brew pub at the Queen's Head in St James Street, Monmouth, which from 1983 urged its customers to get Piston Bitter. In 1985 the old free house was judged in a survey to be selling the cheapest beer in Britain at 54p a pint. For a while the Queen's sold its beers in the free trade under the name Monmouth Fine Ales before ceasing brewing in 1986 after a change of owners.

The most successful and long lasting of the new breweries in Wales were those linked with complementary businesses. Powell's of Newtown had been brewers until 1956, when it closed its Eagle Brewery and concentrated on wholesaling other brewers' beers in Mid-Wales. When the neighbouring Powys Brewery closed, it bought the concern in 1982 and was able to sell its revived Sam Powell beers through its own extensive distribution network, as well as the family's couple of pubs.

In 1988 the wholesale company was sold to Ansells of Birmingham, part of Allied Breweries, and three years later the brewery went out of business when the nine off-licences ran into trouble. However, another small brewery, Woods of Wistanstow, continue to brew Sam Powell's beers for Wales from just over the border in Shropshire.

Pembrokeshire Own Ales near Amroth, founded on a holiday site in 1985, also linked up with a beer wholesaler, Georges of Haverfordwest, to distribute their Benfro beers. But

CHEERS: Peter Johnson of Pembrokeshire Own Ales raises one of the first pints of Benfro Bitter at the brewery on his Llanteglos holiday site in 1985

Georges' parent company, Ansells, took over the running of Georges in 1987 and later dropped the Pembrokeshire brews. Founder Peter Johnson struggled on, delivering the beers himself, until stopping brewing in 1990.

Mr Johnson's Llanteglos holiday complex had at least provided him with one outlet, the club bar, and when experienced brewer Alan Beresford set up his own brewery in 1985, after the closure of North Wales' last major independent brewery, he opted for a building on a farm caravan site, again with a club bar.

His Plassey Brewery literally sprang from the rubble of Border Brewery of Wrexham, taken over and closed by Marstons in 1984. Alan Beresford had been deputy brewer at Border for seventeen years. He rescued some equipment from the scrap merchant and fulfilled a lifetime's ambition by starting his own brewery in an old dairy on a friend's farm at Eyton in the Dee Valley, supplying Farmhouse Bitter to the Treetops club on the farm's caravan site.

Alan Beresford died four years later but his partner Tony Brookshaw has continued the Plassey Brewery with the help of Ian Dale of the Wrexham Lager Brewery. The remarkable Edwardian farm buildings – which won a conservation award – also house a variety of craft workshops and a restaurant alongside the brewery. Plassey is still brewing in 1993, as is one of the most distinctive of the new breweries, Bullmastiff.

Bob Jenkins switched from pintas to pints in 1987 when he bought the brewing plant from Monmouth Fine Ales, the vessels moving in convoy from Monmouth to Penarth near Cardiff on three milk-floats. Previously he had been a keen home-brewer, setting 20-gallon mashes going in his garage before heading off on his milk round.

He named the brewery, set up in an industrial unit in Penarth's old docks, after his prize dogs, and found a willing customer in Will Thomas of the Royal Hotel in Penarth. Soon others followed, especially after his strong ale grabbed the headlines with its tongue-in-cheek title, Son of a Bitch. It was named by a friend the morning after he had enjoyed the night before on the powerful brew.

In 1992 the company, run by Bob and his brother Paul, moved to larger premises in Grangetown, Cardiff, brewing over twenty barrels a week for a widespread trade that extends through wholesalers across England.

Like a few of his predecessors, Bob Jenkins has tried to carve out his own speciality market, rather than just compete head on with the major brewers in the session bitter arena, with his Bullmastiff Bitter and Best Bitter. His SoB, the

NO ARGUMENT: Bullmastiff beer is best, say Bob Jenkins and mascot Tug. Anyone dare disagree? (Western Mail)

strongest draught beer at present brewed in Wales at 6.4 per cent alcohol, is always guaranteed to create a stir, but his Ebony Dark, a porter-like brew, is equally distinctive.

Two of the more enterprising new brewers, Gwent Ales and Pembrokeshire, also tried out different brews. Besides strong ales, they experimented with cask lagers, and Gwent enjoyed some success with Bailey's Stout.

The splendidly named Raisdale Sparging & Brewing Company of the Raisdale Hotel, Penarth, went furthest down this speciality path, with a range of swing-top bottled beers like Looby's Lust, O'Hooligan's Revolt and Stanley's Steamhammer. Sadly, they were not available to the general public, being sold only to hotel residents. But following the sale of the Raisdale, Stephen Simpson-Wells is looking for a site with a full licence.

With a fiercer and more restricted free trade than other parts of Britain, perhaps future new brewers in Wales need to buy a public house first. So far the revival of Wales's home-brew pub tradition has only briefly flickered into life, though in 1992 it received a boost when a new venture appeared in the driest part of the country, the north-west.

Landlord Martin Barry of the Bryn Arms at Gellilydan near Blaenau Ffestiniog decided to brew his own 99p pint because of the rising cost of beer. He called his value-for-money drink Mel-y-Moelwyn – Honey of the Mountains. The tiny brewing plant can be seen through windows in the pub.

In addition, two new free-trade breweries are planned for north-east Wales in 1993. The pint-size revolution has not yet stopped fermenting.

New Brewers

AFAN Brewery, Cwmafan, Port Talbot
(1982–4)

BRAGDY'R DEFAID DU, Henllan,
Llandysul (1980–2)

BRECON Brewery, Camden Arms, Brecon
(1979)

* BULLMASTIFF Brewery, Penarth / Cardiff
(1987–)

CESTRIAN Brewers, Buckley, Mold
(1981–3)

GWENT ALES (Silverthornes), Little
Mill, Usk (1981–5)

GWYNEDD Brewers, Gaerwen, Anglesey
(1980–4)

MINERA Brewery, City Arms, Minera,
Wrexham (1983–9)

MONMOUTH Fine Ales, Queen's Head,
Monmouth (1983–6)

PEMBROKESHIRE Own Ales, Llanteg,
Amroth (1985–90)

* PLASSEY Brewery, Eyton, Wrexham
(1985–)

POWELL, Samuel Powell, Newtown
(1983–91)

POWYS Brewery, Newtown (1981–2)

RED KITE Brewery, Glanyrafon,
Aberystwyth (1982–3)

Home-Brew Pubs

* BRYN ARMS, Gellilydan, Blaenau
Ffestiniog (1992–)

GLOBE INN, Fishguard (1981–8)

HERITAGE INN, St Mellons, Cardiff
(1983–6)

MISKIN ARMS, Miskin, Pontyclun
(1976–9)

RAISDALE HOTEL, Penarth (1985–90)

*still brewing in 1993

DIRECTORY OF WELSH BREWERIES

This geographical directory of Welsh commercial breweries is in alphabetical order of towns and villages, with the name of the pre-1974 counties alongside. The names under each brewery are mainly from trade directories, the date being the first time the brewery appears under that name. Home-brew pubs are not included except where they developed wider trade.

ABERAERON, Cardigan
ABERCLYDAN Brewery, Allt Llwyd Farm, Llanon: David Morgan, 1837–79.

ABERAMAN, Glamorgan
ABERGWAWR Brewery, Gwawr Street: William Williams, 1854; Thomas Howell Jones, 1865; Charlotte Jones, 1880; Taliesin James, 1889. Bought by Nells of Cardiff, 1898.

ABERAVON, Glamorgan
AVONSIDE Brewery, High Street: David Longdon, 1848; Henry Jones, 1880–95.

ABERBEEG, Monmouth
WEBBS: see pages 136–7.

ABERDARE, Glamorgan
ABERDARE & Trecynon Brewery – see Trecynon Brewery.
ABERDARE VALLEY Breweries – see Rock Brewery.
BLACK LION Brewery, Monk Street: Arthur Jones, 1865; Alfred Pleace late 1880s; limited company 1890, liquidated 1898; New Black Lion Brewery, 1899, liquidated 1910 with 23 pubs. Bought by Allsopps of Burton, 1912.

GEORGE Brewery, Commercial Street: Thomas Jones, 1855–1909. Put up for sale with 26 pubs. Executors revived brewing and Thomas Jones & Son survived until early 1950s.
GLOUCESTER Brewery, Dean Street: Thomas Hughes, 1871; Williams & Co., 1876–8.

OLD MILL Brewery, High Street: John Lindsey 1848–67. Merged with Rock Brewery, 1870.
ROCK Brewery, High Street: Thomas Williams, 1856. Lindsey & Fisher, 1871; Edward Roberts, 1875; Rhys &

Lewis, 1876; Richard Lewis, 1879; Morgan David, 1880; limited company 1897 under John Lewis with 19 pubs, liquidated 1910; Alfred Pleace, 1911. Bought by Allsopps, 1913, and merged with Black Lion to form Aberdare Valley Breweries. Rock Brewery renamed Town Brewery. Closed 1921.
TRECYNON Brewery, Mill Street: John Jones, 1865; George Verrier Jones, 1885. Aberdare & Trecynon Brewery formed 1888 under Thomas Rees. Auctioned 1911, only 7 pubs sold. Wound up 1914.
WELSH HARP Brewery, Commercial Street: Set up by group licensees under James Beynon, 1865, mainly to sell Burton ales. Soon stopped brewing.

ABERGAVENNY, Monmouth
BRECON ROAD Brewery: Morgan & Facey, 1871; Frank Morgan, 1874; William Jenkins, 1891.
DELAFIELD's Brewery, King's Arms, Nevill Street: see pages 13–14.
ELLIS: William Ellis brewing in Lion Street (1835), Victoria Street (1840s) and Grofield (1849–52).
FACEY & Son, Market Street: see pages 14–15.
OLD Brewery, Lion Street: Morrison

& Co., 1830s; William Morgan & Co., 1840s. Originally the Phoenix Brewery. REYNOLDS & WASE, New Road: 1840s.
VICTORIA Brewery, Baker Street: Williams & Trotter, 1842; Samuel Trotter, 1849, in Monk Street. Moved Baker Street mid-1850s. Susan & Emma Trotter, 1862, when called The Brewery, Abergavenny. Nathaniel Cook, 1868; James Cook bankrupt 1870. Edward Phillips, 1876; James Gough 1884.

ABERSYCHAN, Monmouth
REFORM Brewery, Union Street: Joseph Duffield, 1832 to mid-1840s. Revived by Daniel Seys Davies of neighbouring Lion Hotel, late 1870s, as the Abersychan Brewery. Merged with Westlakes Brewery of Cwmavon, 1933, and renamed Reform Brewery. Taken over by Buchan's Breweries of Rhymney, 1939.

ABERTILLERY
CWMTILLERY Brewery, Cwmtillery: John Chivers, 1880; Joseph Chivers, 1885; Chivers & Sons, 1895; Chivers Brothers, 1910–14.

ABERYSTWYTH, Cardigan
MORGAN, Isaac, High Street: 1880.
ROBERTS: David Roberts, Trefechan: see pages 116–17.

AMLWCH, Anglesey
AMLWCH Port Brewery, Quay Street: in operation in 1780s. Joseph Jones & Co. (also at Bangor) by 1835; Griffith Evans, 1858; Benjamin Roose, 1868; Thomas Paynter Williamson, 1874; Evans & Fanning, late 1880s. Registered as Anglesey Brewery Co. Ltd, 1900, with 10 pubs. Wound up 1904.
ANGLESEY Brewery: see Amlwch Port Brewery.
BOROUGH Brewery: James Owen, 1844.
MEDDANAN Brewery, Llanfechell: Evan Thomas Hughes, by 1868; Evan Williams, 1874 to early 1880s.

PARYS Brewery, Salem Street: Davies & Owen by 1868; Jones & Owen, 1895; John Owen, 1918; David Jones, mid-1920s. Closed 1933.

BAGILLT, Flint
CAMBRIAN Brewery: William Pierce, 1858; and Son, 1880. Bought by Kelsterton Brewery, 1893. Closed 1899. See pages 17–18.

BALA, Merioneth
EVANS: Morris Evans, Llanfor, 1870s.

BANGOR, Caernarvon
BANGOR Brewery: Joseph Jones & Co., 1835 (also at Amlwch); Hugh Price, 1844.
CITY Brewery: John Pritchard, 1844.
FRICKER: Samuel Fricker, Tros-y-Canol, Penrhos Garnedd. Established 1812. Bought by Arthur Blake, 1889; Beresford Browning, 1895. Taken over by Ind Coope of Burton, 1898. Sold to Anglesey Brewery, 1903.
GLANADDA Brewery: Thomas Jones, 1868.
RATHBONE: Thomas Rathbone, Market Street, 1835; Roberts and Rathbone, 1844.
VICTORIA Brewery, Dean Street: John Ellis, 1858.

BARRY, Glamorgan
BARRY Brewery Co. Ltd, Broad Street: registered 1895, Edgar Thatcher managing director. Ceased trading 1902.

BEAUMARIS, Anglesey
PRINCE OF WALES Brewery, Church Street: John Jones, 1868–95.

BETHESDA, Caernarvon
BETHESDA Brewery: Thomas Williams, 1858; W.P. Williams & Co., 1874; John Griffith, 1880 (licensee King's Arms, High Street). Ceased brewing 1890s.

BLAENAVON, Monmouth
CAMBRIAN Brewery, James Street:

John Griffith Williams, 1844 (also manager of gas and waterworks and the town hall); Charles Francis Westlake, 1884. Westlake's Brewery Ltd, 1889. Moved Cwmavon, 1900.
IVOR CASTLE Brewery: Caleb Edmonds, 1871; William Burgoyne, 1884; Richard James, 1891.

BLAINA, Monmouth
GRIFFITHS Brothers, High Street: begun by Christopher Griffiths, licensee of the Rising Sun, by 1862; Griffiths Brothers Ltd registered 1890 with 20 pubs. Taken over by Andrew Buchan of Rhymney, 1929.

BRECON
BRECON Brewery Co., Watton: Richard Jones, 1844; Seymour Seyer, 1864, bankrupt by 1868. Ceased brewing. Revived 1890s by Guy Dobell and John Courtauld. Registered 1901. Taken over by Hereford & Tredegar Brewery 1905 and brewing ceased. Name used by Dafydd Gittins for a brief new brewery in 1979 at the Camden Arms, a home-brew pub upto 1942.

BRECON Old Brewery, Struet: William Winston by 1835; Thomas Jenkins, 1849; David Williams, 1868; Cansick & Evans, 1871; John Evans, 1875; David Powell, late 1890s. Taken over after David Powell's death in 1937 by Anglo-Bavarian Brewery of Somerset Brewing ceased 1939.

CASTLE Brewery, Church Street: Thomas Baskerville Jones, 1835; James Jones, 1884; John Jones, 1891; Mrs Margaret Jones, 1895.

BRIDGEND, Glamorgan
THE BREWERY: Thomas Lewis, 1840; John Lewis, 1865; Colcoch & Co., late 1860s; Robert Henry Stiles, 1871. Taken over by Simonds of Reading, 1938. See page 18.
REES: Wynne Rees, Pen-y-Fai, 1880.

BRYNMAWR, Brecon
BEAUFORT ARMS Brewery, Beaufort: William Hemus, 1858; Ann Hemus, 1880.
BRYNMAWR Brewery: Thomas Jenkins & Co., 1858; R.B. Jones, 1868. Wound up 1869.
KING STREET Brewery: Walter Jones, 1891; Mrs Catherine Jones, 1901.
WILLIAMS: Evan Griffith Williams, Beaufort Buildings, 1844–58. Revived by Thomas and Anthony Starkey by 1880; W.H. Madeley & Sons, 1891.

BRYNMENYN, Glamorgan
ABERGARW Brewery: John Brothers, founded 1884, registered company 1895. Closed 1940.

BUILTH WELLS, Radnor
BUILTH Brewery, Market Street: founded by John Prosser in 1866. David Williams took over around 1890. Bought by Evan Evans Bevan of Neath in 1952 with 33 pubs and brewery closed.

CAERGWRLE, Flint
LASSELL & SHARMAN: Liverpool brewers who moved to Wales in 1860s, William Lassell brewing by 1868; Lassell & Sharman, 1874. Limited company registered 1894. Taken over by Burtonwood Brewery of Warrington in 1945 with 57 pubs.

CAERLEON, Monmouth
HANBURY Brewery, Hanbury Arms: John Sherwood, 1891; Josiah Brown &

Son, 1901; Mrs Phoebe Rowe, 1904; John Sherwood, 1907; merged with Eastern Valleys Brewery Co. Ltd, Pontnewynydd, 1910, under John Sherwood. Taken over by Hancock's of Cardiff and Newport, 1914.

CAERNARFON
RAE: John Rae, Pool Street, 1835.
WILLIAMS: Thomas Williams, Greengate Street, 1835.

CAERPHILLY, Glamorgan
CAERPHILLY Brewery, Castle Street: Thomas Reynolds, 1866, when known as the Bridge Brewery. Bought by Thomas Williams and company registered 1890.
CASTLE Brewery, Nantgarw Road: founded 1889. Merged with Caerphilly Brewery to form Caerphilly and Castle Brewery, 1893. Taken over by Crosswells Cardiff Brewery, 1897, with 18 pubs. Brewing ceased 1900.

CARDIFF, Glamorgan
ALBION Brewery, 259 Bute Street: Thomas Williams, 1850; Philip Williams, 1858; William Steeds, 1893. Taken over by R.W. Miller of Bristol by 1897.
ANTHONY Birrell, Pearce & Co., Clare Road: Cardiff-based company formed in 1898 with two breweries – in Wick, near Bridgend, and in Fleur-de-Lis, Monmouthshire. Wound up 1899. Revived as Harding & Co., 1901–5.
BEDFORD Brewery: see Roath Brewery.
BLACK LION Brewery, Wharton Street: Fred Dunkley, 1897–1902. Merged with Roath Brewery.
BRAIN'S, St Mary Street: see pages 86–97.
BUTE Brewery, Whitmore Lane: John Reed, 1850; E.T. Jones, 1855; Joshua Rumsey, 1858.
BUTE DOCK Brewery, Bute Bridge, 243 Bute Street: North & Low, 1855; William Hancock, 1883. See pages 99–101.

CAMBRIAN Brewery, Womanby Street: George Watson, 1848; Mrs Watson & Son, 1866; Dominick Watson, 1875. Taken over by Brain's, 1885.

CANTON CROSS Brewery, 271 Cowbridge Road: Thomas Lewis Glaves, 1860; Thomas Wyndham Glaves, 1878; John Clark, 1884; Pearce & Co., 1893; Canton Cross Brewery Co. Ltd, 1897. Taken over by William Hancock 1904.
CARDIFF Brewery Co. Ltd: see Crown Brewery.
CASTLE Brewery, Frederick Street: John Thomas, 1840; Elizabeth Thomas, 1842 (then called New Brewery). 1862 Thomas family bought the Old Brewery, St Mary Street. Henry Anthony took over the New, renaming it the Castle Brewery. Taken over by William Hancock, 1895, with 12 pubs.
COUNTY Brewery, Crawshay Street, Penarth Road: F.S. Lock, 1889; William Hancock, 1894 (renamed The Brewery). See pages 101–2.
CROSS KEYS Brewery, Severn Road: Charles Cleaves, 1865; George Winchester, 1873; Charles Baily, 1875; Ernest Baily, 1884; Pearce & Son, 1885; Pearce & Co., 1891 – in liquidation 1895. Sold to Starkey, Knight & Ford of Somerset and Devon.
CROSSWELLS: see pages 62–5.
CROWN Brewery, John Street: George Watson, 1875; Cardiff Brewery Co. Ltd, 1885; Francis Soule, 1889–93.

EAGLE Brewery, St John's Square (William Nell): see pages 63–4.

ELY Brewery: see pages 71–85.

FRIENDSHIP Brewery, Homfray Street, Bute Terrace: Oliver Jones, 1875; John Roberts, 1878; Thomas Evans, 1880; James Jones, 1890–3.

HANCOCK'S: see pages 98–120.

IVOR PLACE Brewery, Windsor Road: John Follard, 1884; Yorath & Son (from Newport) 1887–95.

KING'S Brewery, King's Road: George Purnell, 1893; Clark & Co., 1897; G.D. Thomas, 1899; Walpole Brewery Co. Ltd, 1900 – in liquidation 1906; New Walpole Brewery Co. Ltd, 1907. Closed 1908.

MARCHIONESS of BUTE Brewery, Cross Street, Frederick Street: Thomas Jenkins & Son, 1863–1922.

NELL: William Nell, St John's Square. See pages 63–4.

NOEL: see Roath Brewery.

OLD Brewery, St Mary Street: see pages 86–97.

PHOENIX Brewery, Working Street: Henry Green Colman, 1865; John Dowson, 1875; Dowson Brothers, 1878; William Hancock, 1888. Closed 1896.

ROATH Brewery, Bedford Place: Howell Brothers, 1880; Thomas Jenkins, 1884; John Hammett, 1885; Frank Granger (trading as the Bedford Brewery), 1887; S.A.H. Noel, 1893;

Noel & Dunkley, 1903; S.H. Noel, 1911; S.A.H. Noel, 1929; limited company, 1946. Closed 1956.

SHIP Brewery, Millicent Street: William Phillips, 1848; Phillips & Son, 1868; William Phillips junior, 1880, Ship Brewery Co. Ltd, 1890. Taken over by Rhondda Valley Breweries in 1899 with 11 pubs.

SOUTH WALES Brewery, Salisbury Road: John Biggs, 1878; Biggs & Co., 1886; Biggs & Williams, 1887. Taken over by William Hancock in 1889 with 12 pubs. Closed 1892. Briefly revived as Biggs & Co., 1899–1900.

TRINITY STREET Brewery: John Biggs, 1863 (moved to South Wales Brewery by 1878); Clark, Cossens & Co., 1878. John Clark took over Canton Cross Brewery by 1884 and Trinity Street closed.

WALPOLE Brewery: see King's Brewery.

WILLIAMS: William Williams, St Mary Street, 1835–mid-1840s. Moved to Old Brewery by 1848.

CARMARTHEN

CARMARTHEN Brewery, Springside: Charles Norton, 1835; Norton Brothers, 1844. Merged with the Merlin Brewery in 1890 to form Carmarthen United Breweries. See pages 142–4.

MERLIN Brewery, John Street: David Evan Lewis & Son, 1868. Merged with brewery above in 1890. See pages 142–4.

VALE of TOWY Brewery, St Peter's Street: James Griffiths briefly in 1860s. Taken over and closed by Carmarthen Brewery. Revived 1890s by Owen Norton. Closed by 1910.

CHEPSTOW, Monmouth

CREESE: Philip Creese, Nelson Street, 1852–71.

MEREWEATHER: George Mereweather, Gate House, High Street, 1868–76.

PRIORY Brewery, The Priory: built 1873. Powell & King, 1876; Thomas

Perkins & Son, 1880. Taken over by George's of Bristol by 1884.

CHIRK, Denbigh

GLEDRID Brewery, Gledrid: Arthur Hughes, 1868, Richard Jones, 1876.

LODGE Brewery, The Lodge, Weston Rhyn: Moses Edwards, 1868; James and Frederick Edwards, 1880; James Edwards & Son, 1904–26.

CONNAH'S QUAY, Flint

KELSTERTON Brewery, Kelsterton: Edward Bate, 1835; Thomas Bate, 1880; company registered 1890. Taken over by Chester Northgate Brewery, 1899, with 93 pubs, and closed 1904.

COSHESTON, Pembroke

COSHESTON Brewery: David White, 1868; John White, 1880. Ceased brewing by 1890.

COWBRIDGE, Glamorgan

ABERTHIN Brewery, Aberthin: Thomas John, 1880; Mrs Janet John, 1895. Closed by 1906 (David Jenkins also brewed in the village in 1890s).

COWBRIDGE (or Bridge) Brewery, High Street: Lewis Jenkins, 1895; Thomas Morgan and Sons, 1906. Taken over by Bass and closed 1955. See pages 19–21.

OLD Brewery, Malt House Lane: Samuel Howells, 1858; Thomas Spencer, 1884; Hansard Brothers, 1901. See page 19.

VALE of GLAMORGAN Brewery, High Street: Lewis Jenkins, 1868. Taken over by William Hancock of Cardiff in the First World War.

CRICKHOWELL, Brecon

RUMSEY PLACE Brewery, Standard Street: Walter Rumsey, 1884; John Rumsey, 1895. Taken over by the Brecon Brewery by 1900.

CRUMLIN, Monmouth

WESTERN VALLEYS Brewery, Viaduct Road: Edmund Edwards, 1895; David Francis Pritchard, 1900.

Taken over by Buchan's of Rhymney, 1930. See pages 60–1.

CWMAVON, Monmouth

WESTLAKES Brewery: see pages 65–7.

DENBIGH

ANWYL: John Anwyl, Lawnt, 1835–58.

COPPY Brewery, Coppy: William Story, 1858; Edward Story, 1886; Elizabeth Story, 1889; Pryce Story, 1895. Closed mid-1930s.

DENBIGH Brewery, Hall Square: Salisbury Williams, 1844; Edwin Hughes, 1858; Edward Angel, 1874; Mrs Angel, 1895; Angel & Co., 1900–15.

GRAIG Brewery, Beacon's Hill: William Parry, 1874; Margaret Parry, 1895. Closed mid-1930s.

HAWK & BUCKLE, Vale Street: William Williams, 1868–95.

HIGH STREET Brewery: John Armor, 1868; Roberts Brothers, 1895.

DOLGELLAU, Merioneth

CAMBRIAN Brewery, Arran Road: Henry Carpenter, 1890s; North Wales Brewery, 1906; Dolgelley Brewery, 1908. North Wales Brewery revived 1909–14.

EBBW VALE, Monmouth

EBBW VALE Brewery: William Jones, 1914; limited company, 1919. Wound up 1926.

LAMB Brewery, Commercial Street, Briery Hill: John Briscoe, 1891–1901.

EWLOE, Flint

CASTLE HILL Brewery: John Fox. See pages 15–16.

FLEUR-DE-LIS, Monmouth

FLEUR-DE-LIS Brewery: David Anthony, 1852; Mary Anthony, 1876; Henry Anthony Berrill, 1891. Merged with Severnside Brewery, Wick, to form Anthony Berrill, Pearce, 1898. Wound up 1899. Fleur-de-Lis Brewery Co., 1906; Gwent Union Clubs Brewery, 1921. Closed 1928.

GLASBURY, Radnor

LLUNVEY SIDE Brewery, Pontithel: John Williams, 1858; David Bridgewater, 1868; Bothwell & Wilcox, 1880s.

GWAUN-CAE-GURWEN, Glamorgan

ABERNANT Brewery, Cwmgorse: Joseph Rees, 1891; Mrs Ann Rees, 1895; John Rees, 1901. Taken over by William Hancock in 1924 with 14 pubs.

HAVERFORDWEST, Pembroke

BRIDGE STREET Brewery: Esther Davies, 1822; George Green, 1858. Ceased brewing by 1884, though continued as maltster.

CLEDDY Brewery, Bridge Street: Thomas Beynon, 1822; Beynon & Son, 1850; Alfred Beynon, 1858; Jones & Daw, 1871; William Jones, 1880.

MORRIS, Bridge Street: John and Thomas Morris, 1822; Sarah Morris, 1830; Ann Morris, 1835. Ceased brewing by 1850.

OAK Brewery, St Thomas's Green: Thomas Martin, 1880; Isaac Johnson, 1890; James Pugh, 1895; Mrs James Pugh, 1920.

SPRING GARDENS Brewery, Mariners' Square, St Martin's: William Crunn, 1830; Matthew Whittow, 1850; Edmond & Rees, 1871. Ceased brewing by 1880, Thomas James continuing as a wine and spirit merchant.

HOLYHEAD, Anglesey

HOLYHEAD Brewery, Newry Street: Richard Williams, 1858; William Jones, 1868. Ceased brewing late 1870s.

NEW Brewery, Old Road: Thomas Jones, 1828; Joseph Jones, 1835.

OLD Brewery, Old Road: William Bulkeley Jones, 1828–35.

PARK Brewery, Newry Street: Hugh Hughes, 1848; Elizabeth Hughes, 1858; Richard Hughes, 1877. Ceased brewing early 1880s.

PORTHYFELIN Brewery: Owen Jones, 1868–74.

HOLYWELL, Flint

RIVER BANK Brewery, Bagillt Road, Walwen: David Williams, 1828; Bill & Marsden, 1844; William Michell, 1868. Closed by 1888.

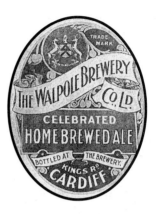

St WINEFRID'S Brewery, Greenfield Street: William Dykins, 1844; Hutchfield & Dykins, 1858; Phillip Dykins & Son, 1868; John Lloyd Price, 1874; St Winefrid's Brewery Company, 1888. Closed 1928. See pages 16–17.

WELL Brewery, Greenfield Street: Price & Littlewood, 1828; Littlewood & Vickers, 1835; Meredith Vickers, 1858; Edward Hutchfield, 1868. Closed by 1880.

LLANASA, Flint

GLANYRAFON Brewery, Glanyrafon: Edward Price, 1858; Alexander Parry, 1868. Ceased brewing by early 1880s.

PROFIT, Glanyrafon: Profit & Williams, 1835; Joseph Profit, 1858. Ceased brewing by 1874.

LLANDILO, Carmarthen

KING'S HEAD Brewery, Bridge Street: William Jones, 1868; Morgan Roderick, 1884. Ceased brewing by 1890.

SOUTH WALES Brewery, Rhosmaen Street: David Lewis founded 1865. Company registered 1888 with 4 pubs. Closed 1912.

LLANDOVERY, Carmarthen

LLANDOVERY Brewery, Stone Street:

James Havard, 1844; John Lewis, 1868; Thomas Watkins, 1880; & Son, 1907. Ceased brewing late 1920s.

ROYAL OAK Brewery, Orchard Street: David Morgan, 1844; Elizabeth Morgan, 1858.

VICTORIA Brewery, Stone Street: Lewis Davies, 1868; Evan Davies, 1880. Ceased brewing by 1910.

LLANELLI, Carmarthen
FELINFOEL Brewery, Felinfoel: David John. See pages 148–60.
LLANELLI Brewery, Thomas Street: Buckleys. See pages 138–47.
NEW Brewery, Station Road: William Bythway. See pages 140–2.
ROYAL EXCHANGE Brewery, Market Street: John Jones, 1865–75.

LLANFACHRAETH, Anglesey
Mona Brewery: John Jones. See pages 28–9.

LLANFAIR CAEREINION, Montgomery
VYRNWY Brewery, Bridge Street: David Jehu, 1889; Jehu Brothers, 1915.

LLANFAIRFECHAN, Caernarvon
RIVER VIEW Brewery, Nantyfelin Road: William Williams, 1889; Sarah Williams, 1895.

LLANFAIR-PG, Anglesey
GARNEDDWEN Brewery: Griffith Roberts, 1858; Morris Roberts, 1868; Griffith Thomas Roberts, 1895.

LLANGEFNI, Anglesey
VICTORIA Brewery, Church Street: Edward Hughes, 1844; William Owen Jones, 1874; Morris Roberts, 1880.

LLANFOIST, Monmouth
LLANFOIST Brewery: James Jones, 1835; Jenkins & Marsden, 1858; Charles Edwards, 1865. Registered 1902. Taken over by Buchan of Rhymney in 1945. See pages 67–8.

LLANGOLLEN, Denbigh
CAMBRIAN Brewery, Regent Street:

Roberts & Davies, 1835; George Roberts, 1844; Ernest Hughes, 1880.
CROWN Brewery, Butler's Hill: Hugh Jones, 1880; John Jones, 1895. Closed by 1912.
HALL STREET Brewery: William Morgan, 1868; Miriam Jones, 1874. Linked with Red Lion Hotel, Bridge Street. Closed by 1909.

LLANGOLLEN Brewery, Berwyn Road: William Morgan, 1835; Walter Booth, 1858; Thomas Booth, 1868; John Tanqueray, 1870. Closed 1919. See pages 20–2.
PRINCE of WALES Brewery, Regent Street: Robert Baker, 1844; George Gault, 1874; Emma Gault, 1880; Ann Baker, 1886; Charles Baker, 1909. Closed by 1918.
SUN Brewery, Regent Street: Elizabeth Jones, 1844. Taken over by Robert Baker of Prince of Wales Brewery by 1868; Mary Baker, 1880; Robert Llewellyn Baker, 1889. Closed 1925.
VICTORIA Brewery, Regent Street: Attwell & Garrett, 1868; William Jones, 1874; Jackson & Co., 1877.

LLANRWST, Denbigh
PLAS-HELIG Brewery, Willow Street: Thomas Williams, 1844; Thomas Elias, 1880. Limited company formed 1905 and wound up the same year.

LLANSANTFFRAID, Montgomery
VIRNIEW Brewery: Kent & England,

1886; Vyrnwy Brewery Co. (Henry Jenks), 1894. Merged with Ellesmere Brewery, Shropshire, to form Ellesmere & Vyrnwy Breweries Ltd, 1897. Closed 1914.

LLANTRISANT, Glamorgan
LLANTRISANT Brewery: Jonathan Thomas, 1875; Taliesin Morgan & Co., 1880–95.
PHOENIX Brewery: Griffith Jenkins, 1891.

LLANTWIT MAJOR, Glamorgan
EAGLE Brewery, Boverton: Rees & David, 1891; George Marriott, 1896; Boverton Brewery Co., 1906. Closed by 1920.
KING'S HEAD Brewery: William John, 1875; Mary Davies, 1891; Alfred Evans, 1895. Ceased brewing by 1906.
WHITE LION Brewery: Alban Watts, 1875; Cecilia Watts, 1895. Ceased brewing by 1906.

MAERDY, Glamorgan
MARDY Brewery: Catherine Davies, 1886; John Davies, 1888. Merged with David John of Pentre, 1888.

MAESTEG, Glamorgan
MAESTEG Brewery, Tywith: Registered December, 1897, by Evan Thomas of the Duffryn Hotel to carry on business of W.J. John and Valentine Pegge with 42 pubs. Taken over by Rogers of Bristol, 1898, and closed.

MERTHYR TYDFIL, Glamorgan
CEFN HILL Brewery, Cefncoed: Catherine Davies, 1875; John Meredith, 1906. Renamed New Brewery by 1910. Closed by 1926.
CEFN VIADUCT Brewery: see Pontycapel Brewery.
CORNELIUS, Quarry Row: David Cornelius, 1835; Mary Cornelius, 1844–52.
CYFARTHFA Brewery, Nantygwenith Street: Herbert Lloyd, 1875; Richard Lloyd, 1906. Taken over by D.T.

Braddick & Son of the Ship Brewery by 1914. Closed mid-1930s.

DOWLAIS OLD Brewery, Dowlais: Charles Powell, 1835; Dowlais Brewery Co., 1848; Charles Sworder, 1868; Rees Atkins, 1875. Ceased brewing by 1880.

GILES & HARRAP: see Merthyr Brewery.

HEOLGERRIG Brewery, Penyrheolgerrig: Evan Evans, 1852; & Son, 1880; Christmas Evans, 1891. Taken over by Pritchard of Crumlin by 1914 and closed.

IRON BRIDGE Brewery, Ynysgaw Street: Thomas Jones, 1865; Lewis Jones, 1871; Owen & George, 1901. Closed early 1920s.

JACKSON BRIDGE Brewery: Williams & Bryant, 1822; William Bryant, 1835.

MERTHYR Brewery, Tydfil Well, Brecon Road: Watkin Davies, 1830; Rowland Hopkins, 1835; Merthyr Tydfil Brewery Co., 1844; James Penny, 1848; John Giles, 1852; Giles & Harrap, 1871. Taken over by William Hancock of Cardiff, 1936. See page 49.

MORGAN, Georgetown: Walter Morgan, 1835–52.

NORTON, Bethesda Street: William Norton, 1852–8.

PENYDARREN Brewery, Penydarren: Thomas Davies, 1875; Penydarren Brewery Co. Ltd, 1892. Closed 1904 after a fire.

PICTON Brewery, Picton Street: John Toop, 1840; Samuel Jones, 1858; Robert Cochrane, 1871; Herbert Cochrane, 1887; Thomas Cochrane, 1891. Ceased brewing by 1894.

PONTYCAPEL Brewery, Cefncoed: Robert Millar, 1848; Thomas Pearce, 1860; Pearce & Shapton, 1871; Harry Pearce, 1880; Pontycapel Brewery Co., 1891; limited company, 1912. Wound up 1921 with 60 pubs. Relaunched as Cefn Viaduct Brewery by 1922 and then New Pontycapel Brewery in 1925. Closed by 1930. See pages 50–2.

SHIP Brewery, Bethel Street: Henry Griffiths, 1858; David Jenkins, 1868; John Morgan, 1871–80. Revived by J. Braddick & Son as Old Ship Brewery by 1895; David Thomas Braddick, 1901. Took over Cyfarthfa Brewery by 1914. Ship closed.

TAFF VALE Brewery, Canal Side: Thomas Evans, 1843; David Williams, 1867; built new brewery above town, 1904; limited company, 1922. Taken over by Buchan's of Rhymney, 1936. See page 50.

TYDFIL Brewery, Tramroad Side: Thomas Davies, 1865; Morgan Morgan, 1871; David Price, 1884. Closed by 1890.

WHEATSHEAF Brewery, Glebeland Street: T. Jones, 1884; Lodovick Mackintosh, 1891; Henrietta Mackintosh, 1894; Emily Botterill, 1897. Closed by 1910.

WILLIAMS: David Williams. See Taff Vale Brewery.

MILFORD HAVEN, Pembroke

MILFORD Brewery, Priory Lodge: Gayer Starbuck, 1844; & Son, 1858; Dayrolles & Williams, 1871. Ceased brewing by 1875. Site used for gasworks.

MILTON, Pembroke

MILTON Brewery, Milton: Thomas Griffiths, 1871; Mrs Thomas Griffiths, 1891; John Jones, 1895; Mrs Elizabeth Jones, 1901. Ceased brewing by 1910.

MOLD, Flint

CATHERALL, High Street: John Catherall, 1835–44.

GLANRAFON Brewery, Wrexham Street: Smith & Price, 1848; Edward Price Jones, 1858; Mary Jones, 1886; Owen Jones, 1895.

OLD MOLD Brewery, New Street: Jones, Lloyd & Co., 1844; David Radcliffe, 1888. Limited company, 1897. Taken over by the West Cheshire Brewery of Tranmere in 1898.

MONMOUTH

MONMOUTH Brewery, Worcester Street: Henry Porter Tippins, 1871; John Alfred Tippins, 1878; Ellis & Walmesley, 1880; Walter Alabaster, 1884; Alan Spurling, 1887; Searle & Co., 1891 (address now St Mary's Street); Vincent & Co., 1898; Henry Porter Tippins, 1901; Harry Westcar Rowland, 1910; Henry Westcar Rowland, 1920. Closed by 1926.

MONTGOMERY

BUCK Brewery, Princes Street: Arthur Withers, 1880; John Withers, 1889; Mary Withers, 1909; Stanley Withers, 1920; S. Richards, 1923.

MOSTYN, Flint

MOSTYN Brewery, Mostyn: John Davies, 1868; Thomas Jones, 1880.

NARBETH, Pembroke

PHILLIPS: David Phillips, St James Street, 1895; Ann Phillips, 1920.

WILLIAMS: James Williams. See Pembroke.

NEATH, Glamorgan

SOMERSET Brewery, Green Street: David Hoskins, 1871; limited company 1874 under Tom Skelton Nash. Closed by 1890.

VALE OF NEATH Brewery, Cadoxton: Evan Evans Bevan. See pages 129–35.

WIND STREET Brewery: Arthur & Eustance, 1848; Benjamin Lanham, 1856. Closed by 1865.

NEWCASTLE EMLYN, Carmarthen
EMLYN Brewery: Daniel Davies, 1868.

NEWPORT, Monmouth
ALEXANDRA Brewery, Constable Lane, Commercial Road: James Herbert, 1880; & Son, 1903; Herbert's Ltd, 1913. Closed and four pubs sold, 1920.
ANCHOR Brewery, Mountjoy Road: William Penny, 1842; Penny & Naish, 1852; Edwin Hibbard, 1858. Taken over by William Hancock of Cardiff, 1884. Closed 1906. See pages 99–100.
BANESWELL Brewery, Baneswell Road: Robert Williams, 1842; Thomas Williams, 1852; Henry Brooks Marriott, 1858. Closed by 1865.

BRISTOL Brewery, Alma Street: Thomas Marshall, 1875; Price & Marshall, 1878; Edgar Thatcher, 1883; Edwin Thatcher, 1891; Thatcher & Son Ltd, 1893. Closed 1931, but revived by another Somerset brewer, A.E. Edwards, in 1933 trading as Thatcher's Bristol Brewery. Bought by Mitchells & Butlers of Birmingham in 1949 and brewing ceased. See pages 127–8.
CAMBRIAN Brewery, Baneswell (later Cambrian Road): John Williams, 1830; Latch & Cope, 1842; Henry Gregory, 1852; W.H. Gregory, 1876; William Yorath & Son, 1877; Lloyd & Yorath Ltd, 1895; Lloyds (Newport) Ltd, 1946. Taken over by Ansells of Birmingham, 1951. See pages 124–7.
CASTLE Brewery, Shaftesbury Street: see pages 121–4.
CLYTHA Brewery, Cannon Street/ Dean Street, Caerleon Road: Richard Williams, 1877; limited company

under Frederick Cross, 1891. Ceased trading late 1890s. Revived by James Gorman by 1902. Ceased brewing 1936.
COMMERCIAL Brewery, West Market Street: John Oxley, 1858; Derrett & Tutton, 1868; Marshall & Horwood, 1871; Robert Derrett, 1875; Prujean & Palmer, 1880; Palmer & Co., 1885; Edwards & George, 1890. Taken over by Daniel Lewis and renamed the Star Brewery, 1892. Closed by 1895.
DOCK ROAD Brewery, East Market Street: John Pearce, 1852; Thomas Floyde Lewis, 1858; Thomas Phillips & Sons, 1874. Taken over by Simonds of Reading in 1949. See pages 124–6.
EAGLE Brewery, Dock Street: William Derrett, 1865; Robert Derrett, 1871. Moved to Commercial Brewery.
HANCOCK: William Hancock. See Anchor Brewery.
LLOYD and YORATH: see Cambrian Brewery.
PHILLIPS & Sons: see Dock Road Brewery.
RAGLAN Brewery, Capel Crescent: George Horwood, 1873–88.
SKINNER STREET Brewery: George Oliver, 1822; Henry Gregory, 1847; Gregory & Evans, 1848. Moved to Cambrian Brewery by 1852.
STAR Brewery: see Commercial Brewery.
THATCHER: see Bristol Brewery.
VICTORIA Brewery, Bridge Street: John Cross, 1868; Isaac Evans, 1877; Mrs Elizabeth Evans, 1882; Ralph Brown, 1887; Mrs Mary Brown, 1892; David Thomas, 1896; Grove Johnson, 1898. Taken over by Crown Brewery, Pontypool, 1900, with six pubs and closed by 1903.

NEWTOWN, Montgomery
ANGEL Brewery, Market Street: Cornelius Morgan, 1874–90.
CAMBRIAN Brewery, New Road: Henry Rowson, 1868; James Carnegie, 1880; Barker Halliwell, 1886. Taken over by Ind Coope of Burton by 1914.

CROWN Brewery, Horse Market: Ray & Davies, 1862; Davies & Issard, 1874; Thomas Issard, 1880; Issard & Dawson, 1886; Montgomeryshire Brewery Co. Ltd, 1890; in liquidation 1893. See page 116.
EAGLE Brewery, Bridge Street: Samuel Powell, 1889; Bert Powell, 1913. Ceased brewing 1956, continuing as wholesalers. Revived brewing under Samuel Powell name, 1983–91. See page 223.
OLD Brewery, Broad Street (Market Street after 1890): Thomas Swift, 1855; & Co., 1922. Taken over by Wrekin Brewery of Wellington, Shropshire, in 1938 with 17 pubs and closed.
VICTORIA Brewery, Broad Street: James Nunn, 1868–85.

PEMBROKE
BURTON Brewery, Dimond Street, Pembroke Dock: J. Llewellyn, 1871; John Meyrick, 1884.
CROMWELL Brewery, Main Street (established 1790 and known as Pembroke Brewery until 1890): Richard Palmer, 1830; Thomas George, 1844; Robert George, 1868; & Son, 1895. Merged with Swansea Old Brewery and Davies' Bonded Stores, Cardigan, 1896. Closed by 1914.
JONES, Pembroke: Thomas Jones, 1850–8.
PEMBROKE DOCK Brewery, Meyrick Street: Henry White, 1858 (began as porter brewer in High Street, opening second brewery in Meyrick Street by 1871); White & Son, 1891; George Cullwick, 1896; limited company, 1897. Wound up 1898. Revival as White & Long ended by 1906.
PEMBROKE STEAM Brewery, Main Street: George Llewellyn Griffiths, 1875. Taken over by wine merchants James Williams of Narberth during the First World War. Ceased brewing by early 1920s.
PHILLIPS: Princes Street, Pembroke Dock: George Phillips, 1844–58.

PRICE: Pembroke Street, Pembroke Dock: William Price, 1850–8.

TREWENT, Pembroke: Trewent & Sons, 1835; William Trewent, 1844–58.

PENARTH, Glamorgan
WINDSOR Brewery, Maughan Street: Fletcher Brothers, 1878; Frederick Lovett, 1884; Lovett & Son, 1885.

PENTRE, Glamorgan
PENTRE Brewery, Llewellyn Street: David John, 1875; limited company 1888, merging David John with John Davies' Mardy Brewery. Taken over by Webbs of Aberbeeg, 1946, with 26 pubs and closed.
LLEWELLYN'S Brewery: Alfred Parfit, 1875; Thomas Lewis, 1884; William Hockaday, 1891. Closed by 1900.

PENYGRAIG, Glamorgan
GREAT WESTERN Brewery, Butchers' Arms: Thomas Morgan Rees, 1881; Mrs Rachel Rees & Son, 1901. Closed by 1910. Revived by Evans, Jenkins & Co.. Closed by 1918.

PONTARDAWE, Glamorgan
SWANSEA VALE Brewery: built as the Pontardawe Brewery for John Jones in 1837; Robert Evans, 1850. Taken over by Vale of Neath Brewery late 1860s and renamed Swansea Vale. Closed 1937.

PONTARDULAIS, Carmarthen
PONTARDULAIS Brewery: Roberts Brothers, 1884; North & Heitzman, 1891. Registered as Cambrian Brewery Co. Ltd, 1893. Closed by 1895.

PONTNEWYNYDD, Monmouth
EASTERN VALLEYS Brewery, Osborne Road: Parkes Brothers, 1849; William Williams, 1858; William Walters, 1862; Levi & Thomas Jones, late 1880s; Percy Jones, 1906. Merged with Hanbury Brewery, Caerleon, 1909, under John Sherwood, as Eastern

Valleys Brewery Co. Ltd. Split up 1912, Pontnewynydd being taken over by Phillips of Newport and closed.

VEATER, Mill Road: James Veater (United Friends Inn), 1901–23.

PONTRHYDYRUN, Monmouth
JENKINS, Terrace Inn: Mrs Eliza Jenkins, 1906; John Jenkins, 1910–14.

PONTYCLUN, Glamorgan
CROWN Brewery, Brynsadler: D. & T. Jenkins Ltd, 1906; United Clubs Brewery, 1919. See pages 199–209.
MORGAN: Thomas Morgan. See Cowbridge.

PONTYGWAITH, Glamorgan
FERN VALE Brewery: David Llewellyn Treharne, 1900; limited company, 1918, in partnership with Thomas Jenkins of the Crown Brewery, Pontyclun. Taken over by Webbs of Aberbeeg, 1949, later becoming part of Welsh Brewers (Bass). Closed 1970.

PONTYPOOL, Monmouth
BAYTON, High Street: Alfred Bayton, 1887; Mrs Ellen Bayton, 1901–20.
CASTLE Brewery: see Pontypool Brewery.

CROWN Brewery, George Street (also known as Three Crowns and Pontypool Steam Brewery): George Hibbell, 1862; Henry Edwards, 1868; Sackville Lupton, 1873; Joseph Harris, 1884; Henry White & Son, 1891; George

Paxton, 1895; limited company, 1896. Taken over by Buchan's of Rhymney, 1902. See pages 58–9.

PONTYPOOL Brewery, George Street: Luce & Thomas, 1830; Morrison & Brough, 1835; Webb & Cadman, 1842; Jane Herbert, 1849; Herbert & Walters, 1852. William Walters left to take over the Eastern Valleys Brewery, Pontnewynydd, by 1862. Mr Lills, 1868; Thomas Neath, 1873; James Gunn, 1890; Donald Reid, 1900. Registered as Castle Brewery Ltd, 1901. Taken over by Westlakes Brewery of Cwmavon, 1911.

PROBYN, Crane Street: John Probyn, 1862; Frederick Probyn, 1895; & Son, 1910; Granville Probyn, 1921.

TROSNANT Brewery, Hanbury Arms, Trosnant Street: John Joshua, 1880; Herbert Joshua, 1906; Elizabeth Joshua, 1920; Harold James, 1923.

PONTYPRIDD, Glamorgan
GLENVIEW Brewery, Court House Street: Henry Hopkins, 1875; New-bridge-Rhondda Brewery Co. Ltd, 1891. Merged with Graig Brewery, 1903, to form Pontypridd United Breweries. Taken over by Rhondda Valley Breweries, 1918. See pages 72–4.

GRAIG Brewery, Rickard Street: David Leyshon, 1867. Merged with Glenview Brewery to form Pontypridd United, 1903. Graig Brewery closed 1904.

PONTYPRIDD Brewery, Taff Street: William Williams, 1854; Pontypridd Brewery Co., 1891. Taken over by Rhondda Valley Brewery, Treherbert, 1892, to form Rhondda Valley Breweries. See pages 72–4.

PONTYPRIDD UNITED Breweries: see Glenview Brewery.

NEWBRIDGE-RHONDDA Brewery: see Glenview Brewery.

STATION STEAM Brewery, Tram Road: John Jabez Evans, 1880; Mrs Anne Evans, 1906.

PORTH, Glamorgan
EIRW Brewery, Eirw Road: John Leyshon, 1880–1904.

PORTHCAWL, Glamorgan
CROWN Brewery, Newton: Henry Jenkin, 1875; Eustace Williams, 1884; Thomas Morgan, 1901; Crown Brewery Co. under Gwilym Morgan John, 1906. Closed by 1914.

PRESTEIGNE, Radnor
BULL Brewery, St David's Street: James Burt, 1884; Mrs Annie Burt, 1891; William Burt, 1895.

PRESTATYN, Flint
OWEN, Tanycoed: Robert Owen, 1874–early 1880s.

PWLLHELI, Caernarfon
VICTORIA Brewery, Carnarvon Road: Samuel Griffith & Son, 1844; Robert Griffith, 1858; Charles Roberts, 1868; Roberts & Thomas, 1880; John Roberts, 1886. Closed early 1890s and converted into a mineral waterworks.

REDBROOK, Monmouth
HALL: James Hall & Nephew, 1842–52.
REDBROOK Brewery: Graterax & Dyke, 1842; Charles Herbert, 1849; Thomas Burgham & Son, 1858; Henry Burgham, 1868; Mrs Eliza Burgham, 1871; Arthur Burgham, 1895. Taken over by Ind Coope of Burton in 1923 with 22 pubs and closed.

RHUDDLAN, Flint
ABBEY Brewery: John Roberts, 1835–58.
HUGHES, High Street: William Hughes, 1844–68.

RHYDYMWYN, Flint
RHYDYMWYN Brewery: John Lloyd, 1868–76.

RHYL, Flint
HIGH STREET Brewery: John Pierce, 1844; Sarah Corell, 1858; C. Jones & Co., 1868.

RHYMNEY, Monmouth
BUCHAN: Andrew Buchan. See pages 55–70.

RISCA, Monmouth
RISCA Brewery: Thomas Cross, 1852; Cross & Matthews, 1858. Taken over by William Hancock, 1902, and closed 1916.

RUTHIN, Denbigh
CORPORATION ARMS Brewery, Castle Street: Robert Roberts, 1876 (took over Hand Brewery by 1894); William Owen, 1895.
HAND Brewery, Well Street (known as Ruthin Brewery until late 1880s): William Edwards, 1874; John Foulkes, 1886; Benjamin Trimmer, 1889; Robert Roberts, 1895. Taken over by Ind Coope of Burton, 1917.
PARK PLACE Brewery, Mwrog Street: William Jones, 1895; Margaret Jones, 1909–20.
SPREAD EAGLE Brewery, Upper Clwyd Street: Edward Evans, 1876; Joseph Morris, 1879; Charles Phillips, 1895.

ST CLEARS, Carmarthen
SANTA CLARA Brewery: George and Benjamin Hughes, 1858; D. & W. Davies, 1868; Davies, Sinclair & Co., 1880. Gwilym Evans, late 1880s. Absorbed into Carmarthen United Breweries, 1890.

SAUNDERSFOOT, Pembroke
SAUNDERSFOOT Brewery: Thomas Stephenson, 1850; William James, 1858; Benjamin Thomas, 1868. Ceased brewing by 1884.

SENNYBRIDGE, Brecon
PONT SENNY Brewery, Sennybridge: George Cave, 1875; George Lane, 1880.

SWANSEA, Glamorgan
BRYNHYFRYD Brewery, Brynhyfryd: Leyshon Matthews, 1875–1901.
CAMBRIAN Brewery, Strand: see pages 47–8.
GLAMORGAN Brewery, Little Madoc Street: Thomas Jones, 1881; Glamorgan Brewery Co., 1889. Merged

with Orange Street Brewery, 1890, to form Swansea United Breweries. See pages 109–10.
HAFOD Brewery, Neath Road: Thomas White, 1865; Hafod Brewery Co. Ltd, 1881; Alfred Price, 1883. Closed by 1886.
HANCOCK: see pages 100–20.
HARRISON, Woodfield Street, Morriston: George Harrison, 1900; Harrison & Co., 1904. Ceased brewing by 1922.
HIGH STREET Brewery, Tower Lane: Hannen & Riley, 1835; James McKnight, 1840; William Ballinger, 1848; William Ballinger junior, 1871; Thomas Jones, 1881. Taken over by William Hancock, 1891, with 13 pubs. Closed 1894.
ORANGE STREET Brewery: John Andrew & Son, 1850; Philip Andrew, 1854; Andrew & Crowhurst, 1865; Henry Crowhurst, 1875; Crowhurst & Windsor, 1887. Merged with Glamorgan Brewery, 1890, to form Swansea United Breweries.
OXFORD Brewery, Oxford Street: Andrew Brothers, 1865; John Evans, 1883.

ST MARY STREET Brewery: Thomas Hanson, 1850; George Bydder, 1858.
SINGLETON Brewery: see Wellington Brewery.
STRAND Brewery, Strand: Edward David, 1822; Thomas Jones, 1835.
SWANSEA OLD Brewery, Singleton Street: see pages 105–10.

SWANSEA UNITED Breweries: see pages 109–10.

WELLINGTON Brewery, Little Gam Street: David Wechio, 1852; Thomas White, 1858; Joseph Clark, 1865; Clark & Son, 1884. Revived as Singleton Brewery by Moseley & Jarvis, 1890; David Jarvis, 1895; Hoddinott & Mears, 1899; Singleton Brewery Co. Ltd, 1899. Taken over by William Hancock, 1917.

WEST END Brewery, Little Madoc Street: John Ackland, 1865; Mrs Elizabeth Ackland, 1876; Ackland & Thomas, 1883. Taken over by William Hancock, 1890. See pages 101–20.

TALGARTH, Brecon
TALGARTH Brewery, Bronllys: Davies Brothers, 1907; Ernest Walter Davies, 1914; limited company, 1924. Ceased brewing 1935.

TENBY, Pembroke
DAVIES, Creswell Street: James Davies, 1844–58.
SMITH, St Julian Street: John Smith, 1844–75.
TENBY Brewery: Thomas Dundridge & Co., 1875; William Dunning, 1877.

TON PENTRE, Glamorgan
LAVENDER: William Lavender, 1875–80.

TONYPANDY, Glamorgan

TONYPANDY Brewery, Dunraven Arms Hotel: Limited company formed 1885. Taken over by Crosswells of Cardiff, 1899, and closed.

TREDEGAR, Monmouth
TREDEGAR Brewery, Church Street: William Robins, 1835; Joseph Robins, 1865; Joseph and Richard Robins, 1868; Richard Jenkins & Son, 1884; John Jenkins, 1891. Merged with George Edwards of Hereford, 1899, to form the Tredegar & Hereford Brewery Company. See page 67.

TREHARRIS, Glamorgan
TREHARRIS Brewery: Thomas Lewis Williams, 1884; private company formed 1886. Ceased trading 1900.

TREHERBERT, Glamorgan
RHONDDA VALLEY Brewery: see pages 72–3.

TREORCHY, Glamorgan
PENCELLI Brewery, Cardiff Arms Hotel: John Lewis, 1884; Mrs Mary Lewis, 1891.

WELSHPOOL, Montgomery
BROOKSIDE Brewery, Back Road: Richard Rider, 1858; Thomas Rider & Sons, 1868; Rider Brothers, 1874. Ceased brewing late 1880s, the premises being converted into a mineral water factory.

WENVOE, Glamorgan
WENVOE Brewery: registered 1902 and liquidated 1903.

WICK, Glamorgan
BROUGHTON COURT Brewery: John Jabez Evans, 1895; Lewis Harry, 1906; Mrs Jane Harry, 1920. Still brewing in 1926.
SEVERN SIDE Brewery: Margaret Morgan & Sons, 1891; William Herbert, 1895. Merged with Fleur de Lys Brewery, Monmouthshire, 1898, in new company Anthony Birrell, Pierce, based in Cardiff. Liquidated 1899.

WREXHAM, Denbigh
ALBION Brewery, Town Hill: Edward Thomas, 1799; Edward Crewe, 1818; John Williams, 1835; William Williams, 1850; John Beirne, 1868. Closed and 23 pubs bought by Wrexham Lager Brewery, 1938.
BORDER Breweries: see pages 176–83.
BRIDGE HOUSE Brewery, Wrexham Fechan: William Eyton, 1868; Hannah Eyton, 1879. Merged with Sun Brewery, 1880.
BRIDGE STREET Brewery: John Beardsworth, 1835; Dicken & Beardsworth, 1850; John Dicken, 1858.
BURTON Brewery, Bridge Street: Edward Evans, 1822; Evans & Son, 1835; Charles Evans, 1858; Julius Chadwick, 1875. Closed 1922, pubs sold to F.W. Soames.
CAMBRIAN Brewery, Bridge Street: Clark & Orford, 1844; Joseph Clark, 1850; Clark & Sons, 1868; William Sisson, 1874. Closed 1922, leasing 48 pubs to Island Green Brewery.
EAGLE Brewery, Bridge Street: Peter Williams, 1850; Thomas Williams, 1858; T.R. & J. Williams, 1868; Robert Williams, 1874. Merged with Sun Brewery by 1886 under Thompson & Co.; Heasman & Co., 1909. Closed 1922.
ISLAND GREEN Brewery: see pages 176–8.
MANLEY, Hope Street: Thomas Manley, 1868. Took over Vicarage Hill Brewery, 1880.
MITRE Brewery, Pentrefelin: David Price, 1868 (and Town Hill Brewery, 1874); Martha Price, 1886; Margaret Price, 1895. Closed in First World War.
NAG'S HEAD Brewery: see pages 170–5.
OLD SWAN Brewery, Abbot Street: Richard Lovatt, 1835; Edward Lovatt, 1868–80.
PENTREFELIN Brewery, Pentrefelin: Kyffin & Jones, 1876; S. & T. Jones, 1877; Davies & Jones, 1880; Joseph Davies, 1886.

SOAMES: see pages 170–5.

SUN Brewery, Abbot Street: Joseph Craven, 1858; James Thomas, 1868; Rowland & Co., 1874. Merged with Bridge House Brewery under Guirron, Parry & Thomson by 1880. Merged with Eagle Brewery by 1886.

THREE TUNS Brewery, Town Hill: Joseph Jones, 1822; James Smith, 1874.

TOWN HILL Brewery: John Rowland, 1858; George Rowland, 1868. Merged with Mitre Brewery by 1874.

UNION Brewery, Tuttle Street: Charles Bate, 1840; Bate & Son, 1874; George Bate, 1886. Taken over by Peter Walker of Warrington, 1909.

VICARAGE HILL Brewery: William Jones, 1844; Ann Jones, 1850; Jones Brothers & Goodwin, 1868; Rowland & Edwards, 1878; Cheetham & Edwards, 1879; Thomas Manley, 1880. Closed by 1894.

VICTORIA Brewery, Holt Street:

J. Morley, 1873. Liquidated, 1878.

WALKER: See pages 167–9.

WELL Brewery, Brook Street; John Vaughan, 1850; Price Vaughan, 1858–80.

WILLOW Brewery: see Walker.

WREXHAM Brewery: see page 170.

WREXHAM LAGER Brewery: see pages 184–98.

WYNNSTAY Brewery, York Street: John Murless, 1880; Murless & Loxham, 1884; Murless & Co., 1889. Ceased brewing by 1894. Later used as stores by Bents Brewery, Liverpool.

YSTRAD, Glamorgan

YSTRAD Brewery: John James, 1884.

Labels courtesy of Keith Osborne

INDEX

This is only an index of the main text and not of the geographical directory of Welsh breweries.